IMAGES
of America

David Sarnoff Research Center

RCA Labs to Sarnoff Corporation

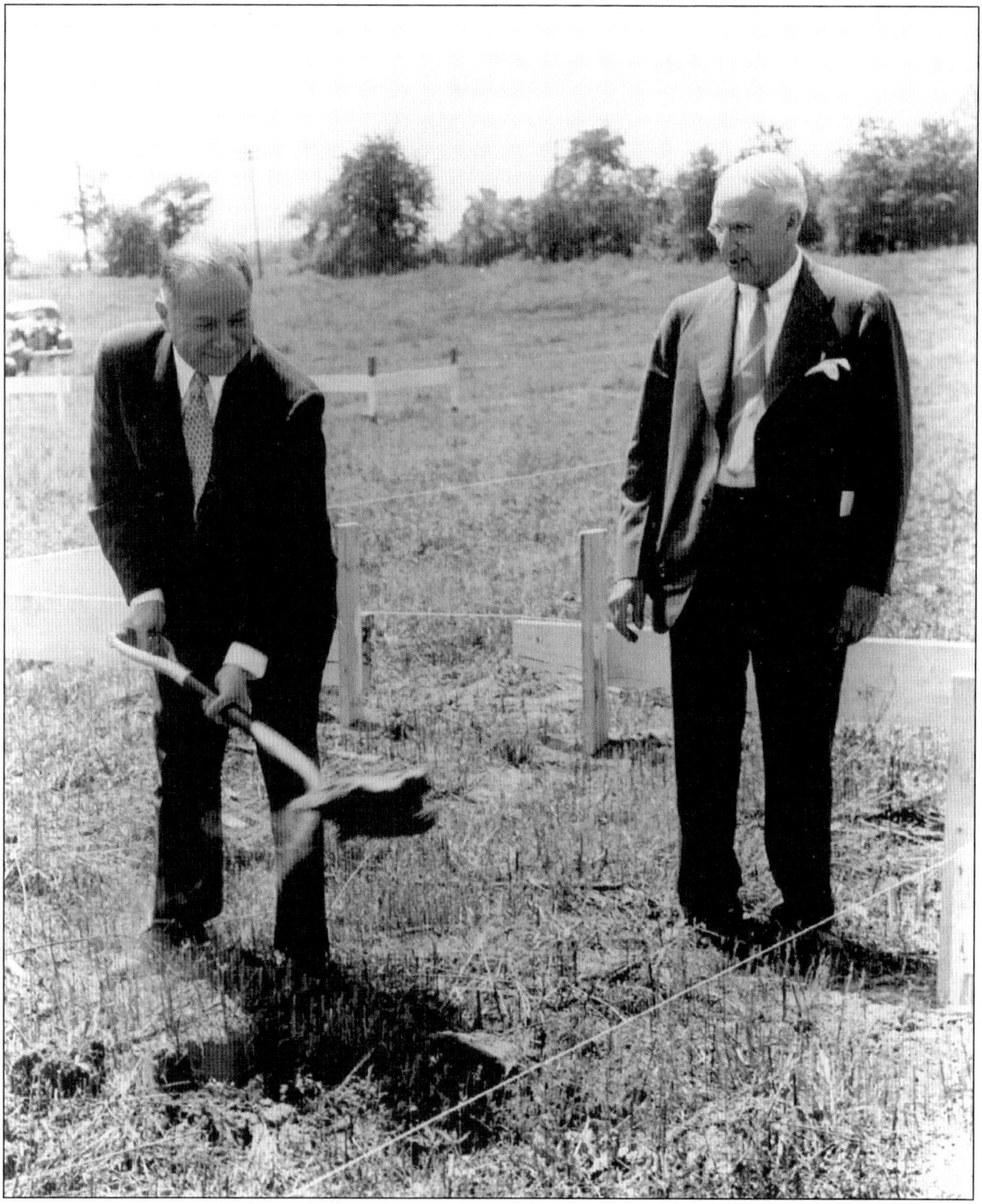

On August 8, 1941, RCA president David Sarnoff breaks ground for the new $1 million RCA Laboratories as Otto Schairer, vice president for RCA Labs, looks on.

On the cover: In a 1951 photograph, Eve Lynn and Joyce Morton are ecstatic about the prototype electronic color television set invented at RCA Laboratories and built by RCA Victor in Camden, New Jersey.

IMAGES
of America

David Sarnoff Research Center

RCA Labs to Sarnoff Corporation

Alexander B. Magoun

ARCADIA
PUBLISHING

ISBN 978-0-7385-1331-8

Published by Arcadia Publishing
Charleston, South Carolina

Printed in the United States of America

Library of Congress Catalog Card Number: 2003108748

For all general information contact Arcadia Publishing at:
Telephone 843-853-2070
Fax 843-853-0044
E-mail sales@arcadiapublishing.com
For customer service and orders:
Toll-Free 1-888-313-2665

Visit us on the Internet at www.arcadiapublishing.com

Dedicated to the late M. Phyllis Smith of Sarnoff Corporation for her lifelong devotion to the Princeton laboratories in all their incarnations.

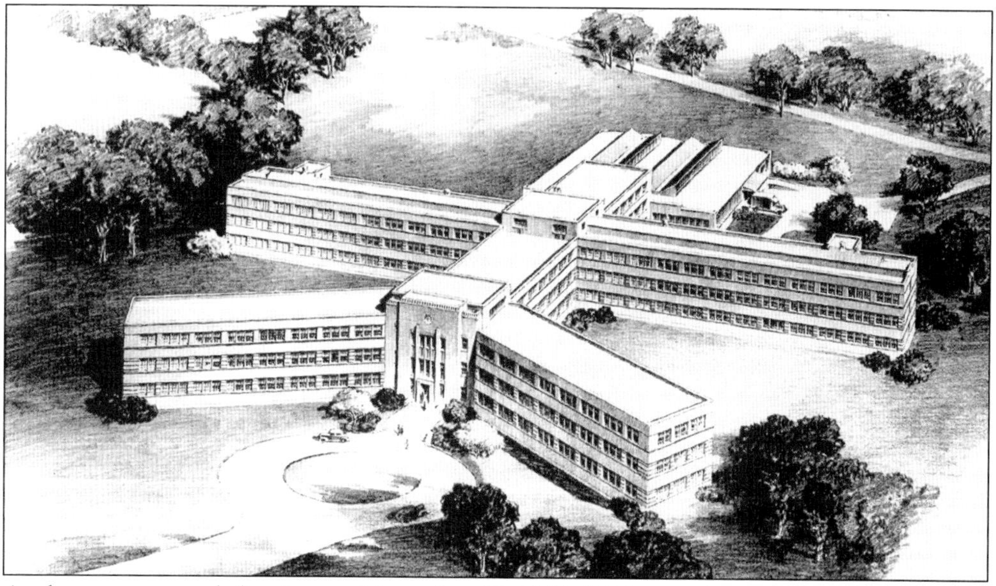

At the cornerstone dedication on November 15, 1941, RCA revealed this final design for the labs by architect Harry L. Porter. A grand entrance connects to the Lincoln Highway (Route 1), and a right of way leads to the Millstone River Sheep Wash.

Contents

ACKNOWLEDGMENTS

I want to thank the following people and organizations without whose interest in, support of, and dedication to history and innovation this book would not be possible: the staff photographers of RCA Laboratories and the David Sarnoff Research Center (DSRC)—Norman Newell, Thomas Cook, and Martin Zak—many of whose compositions grace these pages; the many former and current employees of the DSRC and Sarnoff Corporation who provided information; Thomson/RCA; the New Jersey State Library and the New Jersey Historical Commission; archivists Richard Trotter and Karen Siracusa, who organized and cataloged the DSRC photograph collections; Andrew Matey and Byron Chen of the Hamilton and Edison, New Jersey, Boy Scout troops, respectively, whose Eagle Scout projects digitized the RCA Labs' photograph catalogs; the library, creative services, and marketing staff at Sarnoff Corporation; the board of directors of the David Sarnoff Collection; and Satyam Cherukuri, president and chief executive officer of Sarnoff Corporation, whose appreciation for past accomplishments is surpassed only by his dedication to the company's current innovations and his vision for its future.

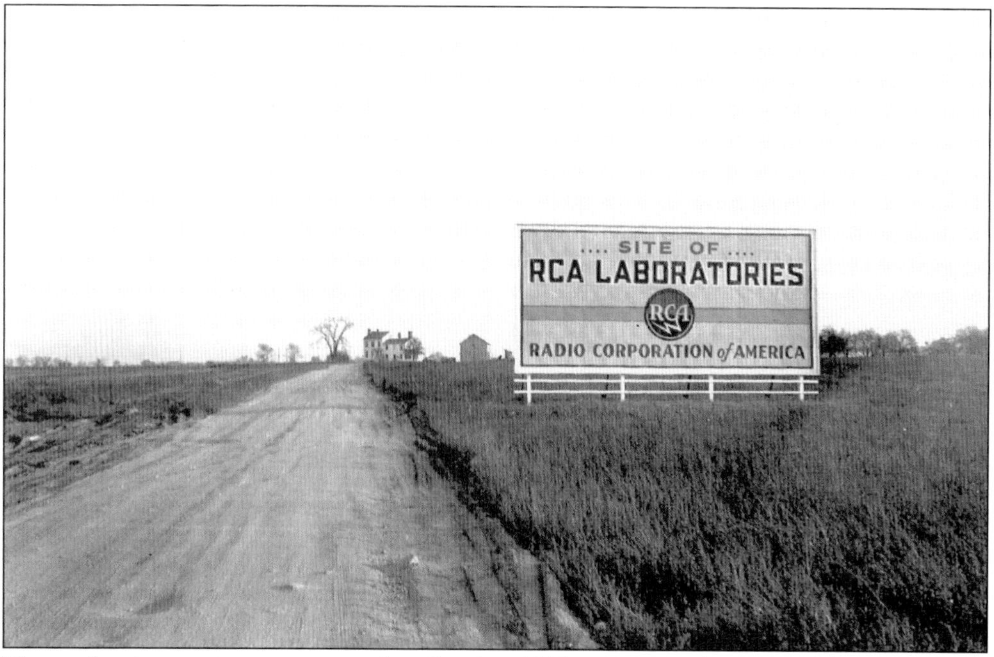

The road to innovation was unpaved in October 1941, as seen in this view of the Route 1 entrance to the future laboratories.

INTRODUCTION

This is the history of an institution whose name has changed but whose mission, after 60 years, remains the same: innovation. The David Sarnoff Research Center has always been about more than invention—the idea that is "new, useful, and not obvious to one skilled in the art," as U.S. patent law defines the term. Innovation means taking that idea through the steps of demonstration, practice, prototype, production model, and sales to a new market. In other words, the David Sarnoff Research Center has always embodied the mindset of its namesake, who wrote in 1915, "I have in mind . . . a Radio Music Box" and proceeded for the rest of his career to make a vibrant reality of that and other industries.

How innovation comes about and who pays for its satisfactions and successes, as well as its frustrations and failures, varies over time. In this respect, the experiences of the center represent those of American industry as a whole. Innovation, moreover, defines our economy, our society, and our culture. We could not be Americans without assuming that we can make better ways to catch mice, to communicate, to stay healthy, and, ultimately, to live better lives. The trick lies in predicting which new mousetrap the consumer will buy. RCA Laboratories—later the David Sarnoff Research Center and now Sarnoff Corporation—enjoy a reputation for making technologies that work, and work better than their alternatives. More than the color television system that we still enjoy, these technologies include cameras and computer vision, lasers and liquid-crystal displays (LCDs), microscopes and microwaves, satellites and semiconductors.

There is no one best route to successful innovation, but David Sarnoff devoted his life to ensuring that the Radio Corporation of America (RCA) covered all of its bets as the most innovative company of the 20th century. The RCA Labs played a central role in his strategy as a facility for independent and government-funded research, development from breadboard invention to prototype, patent management and licensing, and production coordinated with RCA's divisions. As a result, for more than two generations, we have lived in a world surrounded by Sarnoff technologies.

Had he lived long enough, Sarnoff would have championed the global information network that he predicted 40 years ago and that we all enjoy today. It turns out that the men and women of the David Sarnoff Research Center have played a central role in making it possible. Their efforts did not occur in a vacuum, for they built on the work of others while the public, the government, and other companies bought the center's technologies and contributed to their further development and success. For that reason, and because their job is to make new things work, the members of the staff have seldom considered their work to be heroic. But being creative, and sometimes graceful, in creating a vital new tool or technology under the pressure of military or commercial urgency may merit that description. The history of accomplishments by Thomas Edison is well known; the history of innovative creativity at the low-slung building looking over Route 1 is less so. This book is a small effort to acknowledge the work done there by many talented individuals and teams throughout the organization's history.

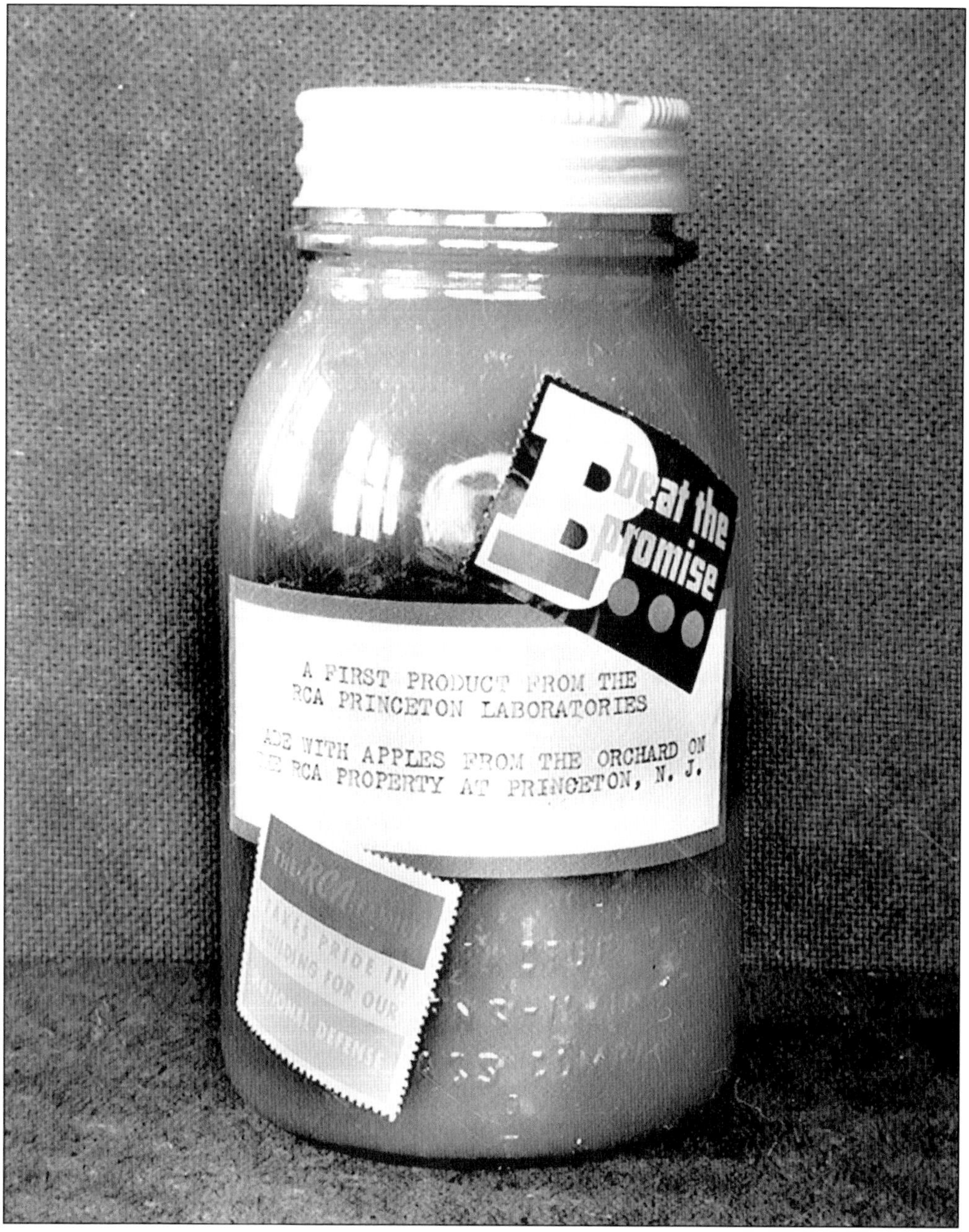

As the staff arrived in the fall of 1942, they drew on local resources to help RCA and the nation "Beat the Promise" of increased wartime production. Ed Dickey led the harvest from the orchard where the west wing extension now stands, resulting in the first product of RCA's most modern facility: apple sauce. One RCA executive praised his jar of apple sauce as "establishing high standards of quality which we hope may be attained in future products in other fields."

One

From Farm to Physics 1941–1945

RCA Laboratories in Princeton originated from a blend of pressures on the parent company at the end of the 1930s. Research and development of vacuum, or electron, tubes took place at the RCA Radiotron Division factory in Harrison, New Jersey. Research and development for products using those tubes took place at the RCA Victor Division in Camden. The facilities were old and the red tape of coordination enormous. Coincidentally, RCA was contracting for more research and development and production for the armed forces in 1940–1941. A separate location for innovation would free up factory space, unify research and development, increase security against espionage, and ease planning for future innovations.

Thus, a year to the day before Pearl Harbor, two RCA executives toured Sarah Olden's farm in West Windsor, which the company purchased in March 1941. This spot was centrally located between RCA's factories and a short drive or rail trip to the company's Rockefeller Center headquarters. Princeton was also the site of the eponymous university and the Institute of Advanced Study, whose leaders welcomed the new institution. For the duration, however, the research staff of RCA Labs focused on high-tech innovations that would win the war more efficiently.

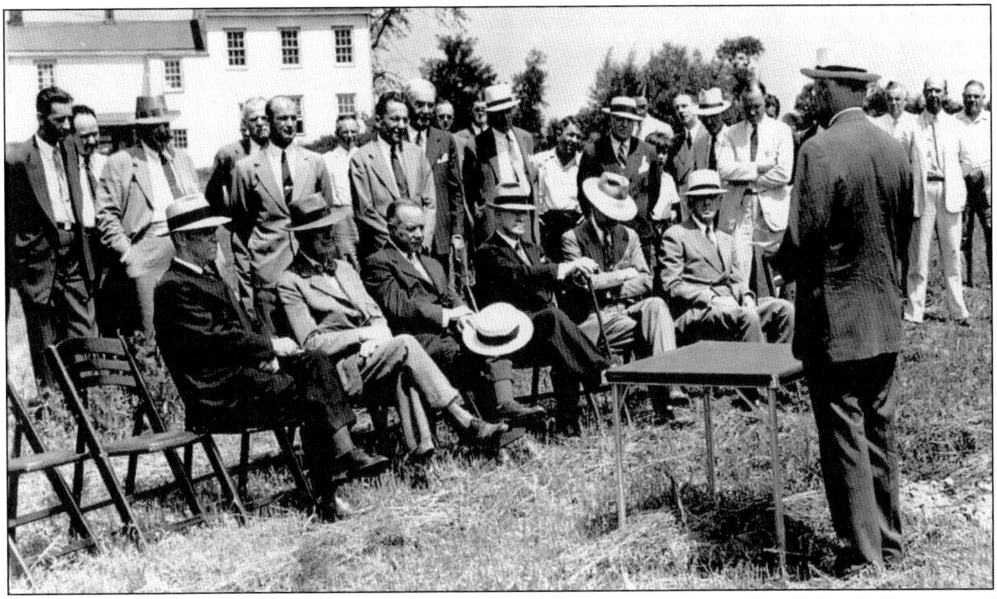

Under a blazing sun, a group listens to an unidentified speaker at the RCA Laboratories' groundbreaking ceremony on August 8, 1941. Gen. James G. Harbord, RCA chairman, and Gen. David Sarnoff, RCA president, are seated second and third from left, respectively.

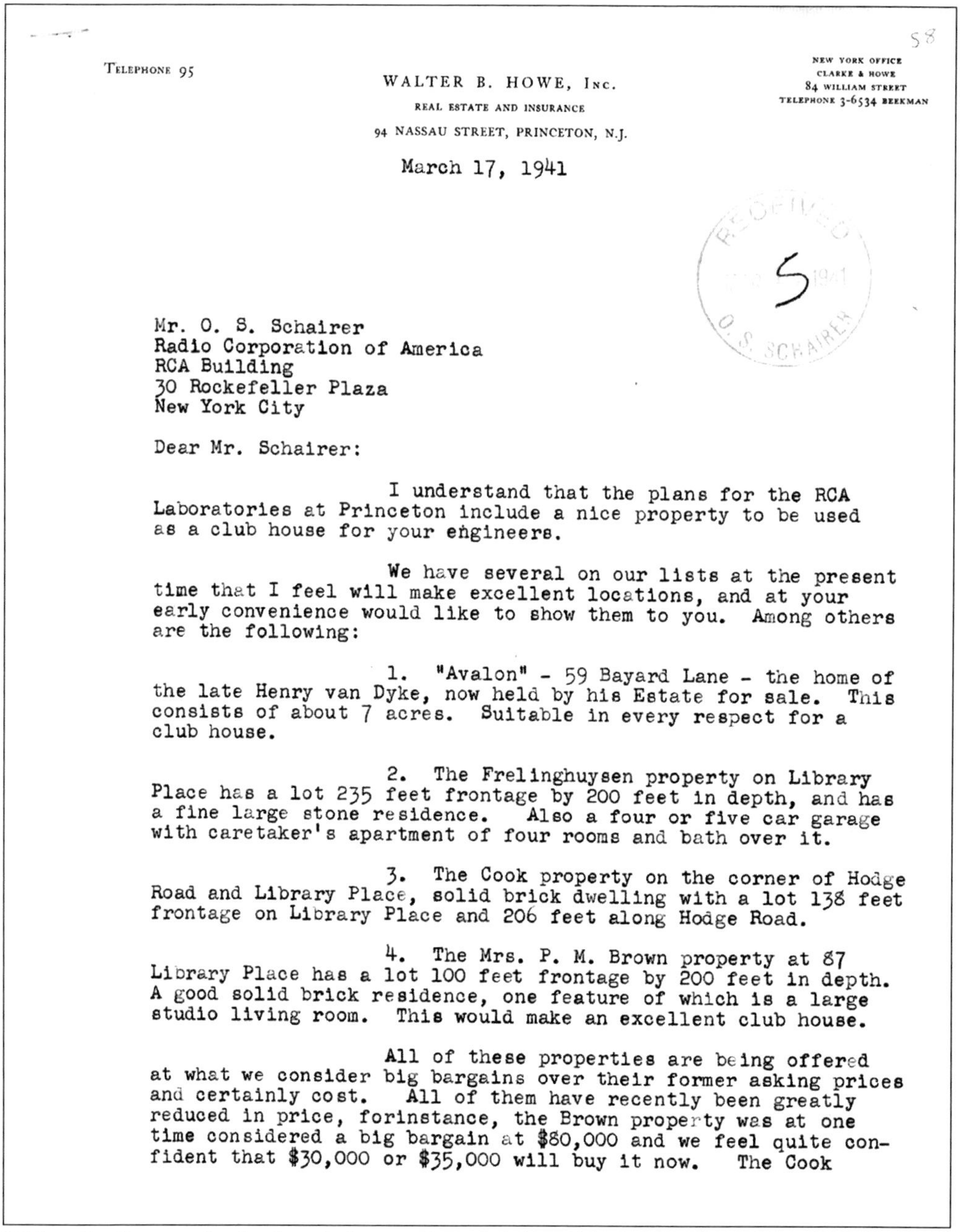

TELEPHONE 95

WALTER B. HOWE, INC.
REAL ESTATE AND INSURANCE
94 NASSAU STREET, PRINCETON, N.J.

NEW YORK OFFICE
CLARKE & HOWE
84 WILLIAM STREET
TELEPHONE 3-6534 BEEKMAN

March 17, 1941

Mr. O. S. Schairer
Radio Corporation of America
RCA Building
30 Rockefeller Plaza
New York City

Dear Mr. Schairer:

I understand that the plans for the RCA Laboratories at Princeton include a nice property to be used as a club house for your engineers.

We have several on our lists at the present time that I feel will make excellent locations, and at your early convenience would like to show them to you. Among others are the following:

1. "Avalon" - 59 Bayard Lane - the home of the late Henry van Dyke, now held by his Estate for sale. This consists of about 7 acres. Suitable in every respect for a club house.

2. The Frelinghuysen property on Library Place has a lot 235 feet frontage by 200 feet in depth, and has a fine large stone residence. Also a four or five car garage with caretaker's apartment of four rooms and bath over it.

3. The Cook property on the corner of Hodge Road and Library Place, solid brick dwelling with a lot 138 feet frontage on Library Place and 206 feet along Hodge Road.

4. The Mrs. P. M. Brown property at 87 Library Place has a lot 100 feet frontage by 200 feet in depth. A good solid brick residence, one feature of which is a large studio living room. This would make an excellent club house.

All of these properties are being offered at what we consider big bargains over their former asking prices and certainly cost. All of them have recently been greatly reduced in price, forinstance, the Brown property was at one time considered a big bargain at $80,000 and we feel quite confident that $30,000 or $35,000 will buy it now. The Cook

Walter B. Howe, who brokered the purchase of the Olden Farm for RCA, also had some suggestions for a suitable clubhouse for the staff's recreation after hours. Instead, the staff frequented the Penn's Neck Community Club. To calculate the equivalent 2002 prices for Howe's recommendations, multiply his figures by 12.

RCA vice president Otto Schairer, who championed the establishment of a separate facility for RCA's centralized innovation operations, digs in with brio at the groundbreaking on August 8, 1941.

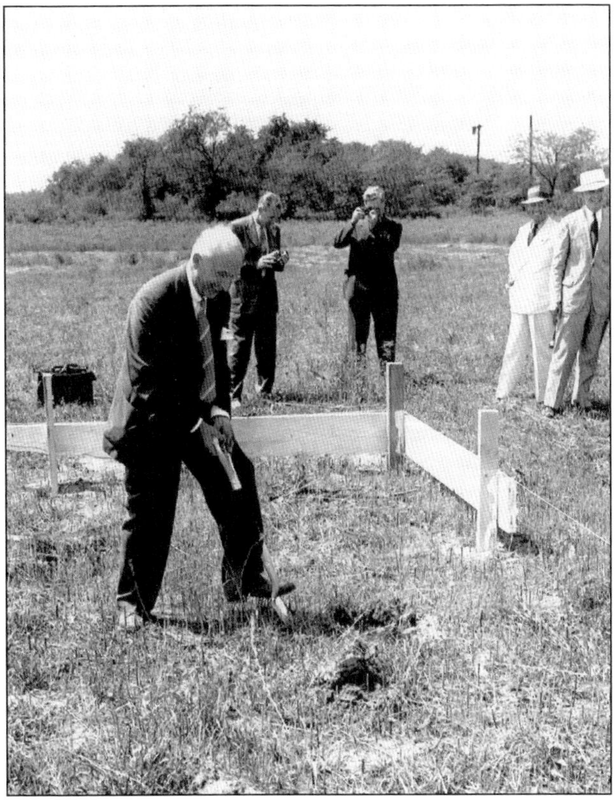

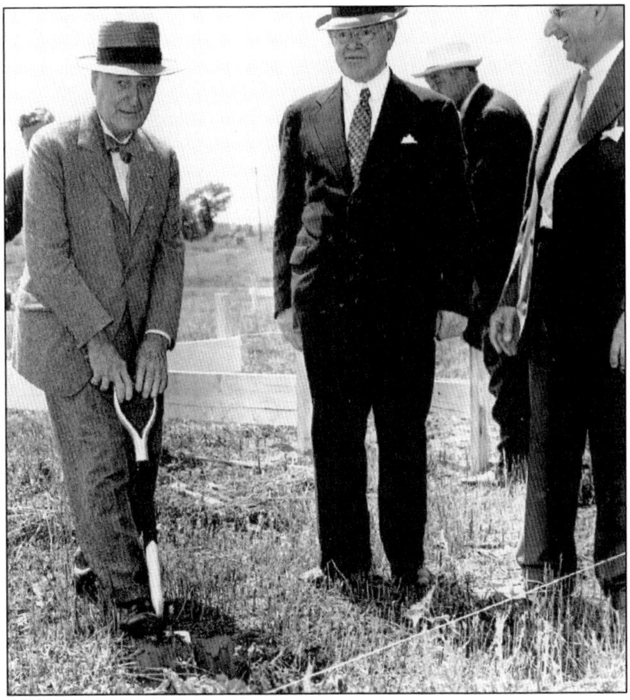

Former RCA presidents—RCA chairman James G. Harbord (1923–1929), on the left, and RCA board member Edward J. Nally (1919–1923)—join in the occasion.

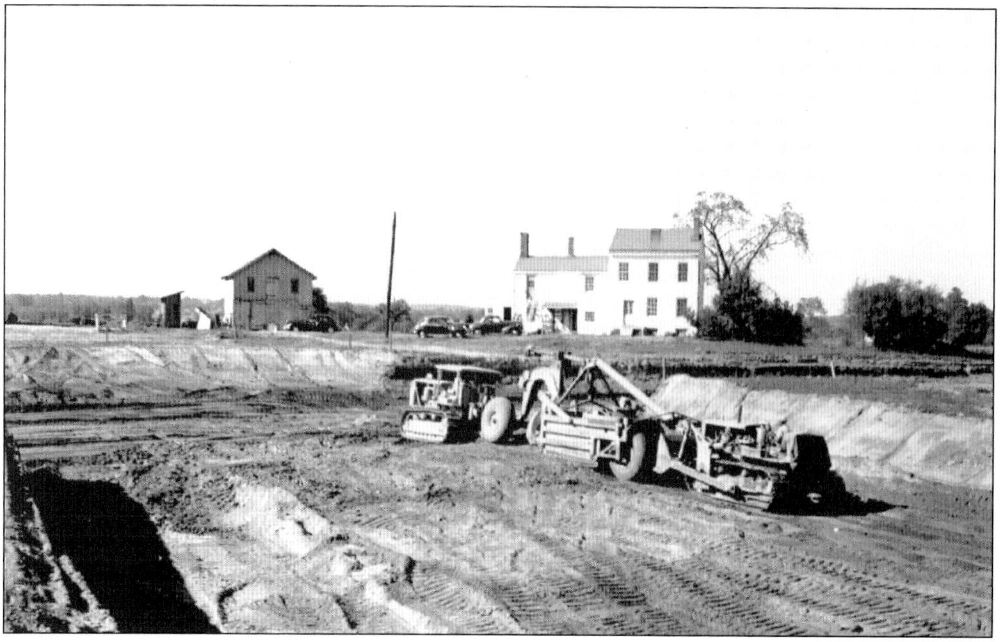

Excavations for the laboratories begin immediately after the groundbreaking in August 1941.

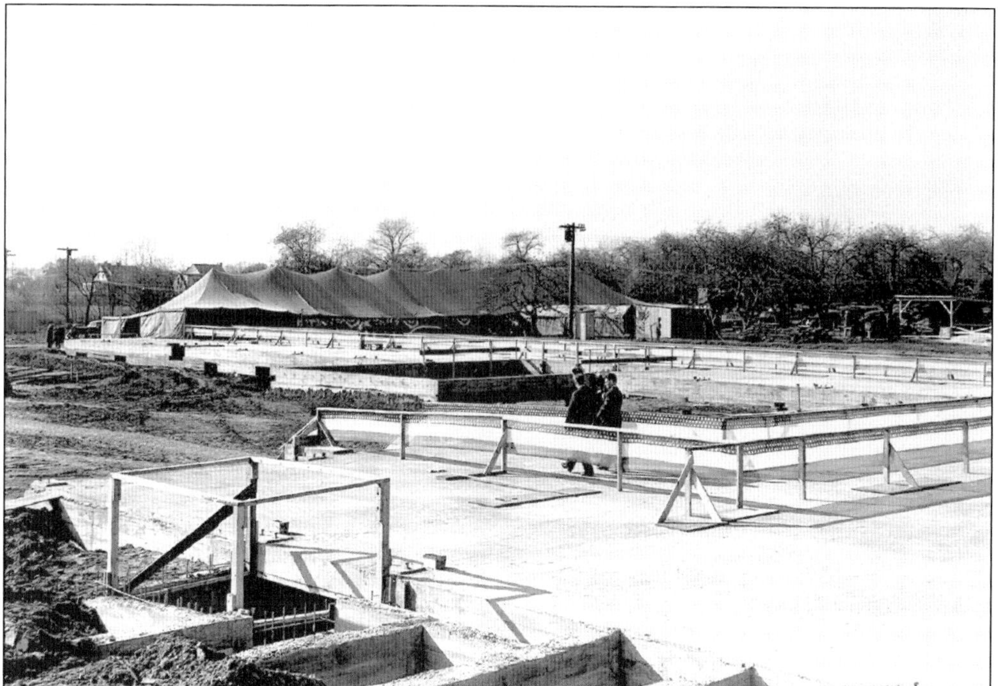

The cornerstone tent at the end of the west wing is seen on November 15, 1941, from the model shop at the base of the south wing. The date was chosen in part as the 15th anniversary for the National Broadcasting Company. Note the Stars and Stripes decorating the walkway, and the apple orchard where the west wing extension now stands.

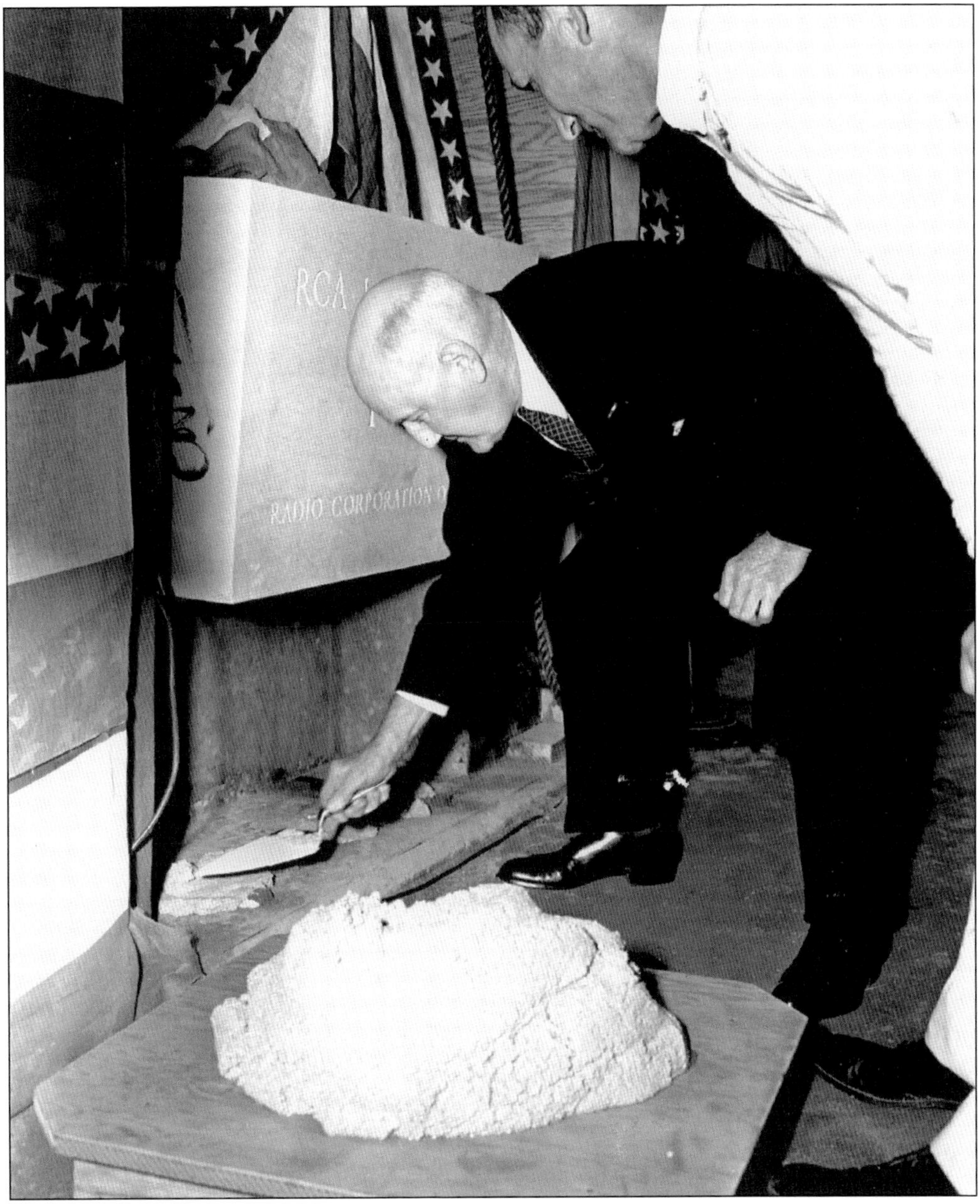

Gen. James G. Harbord, RCA chairman, lays cement ceremonially for the cornerstone, located at the end of the center's west wing. RCA president David Sarnoff, speaking from the *Matsonia* en route to Pearl Harbor, pledged "the unremitting efforts of RCA Laboratories to the national defense of the United States." Harbord echoed his comments and looked further, concluding, "these Laboratories shall serve greatly in the defense of our country, and in the rebuilding of a stronger nation in the years to come."

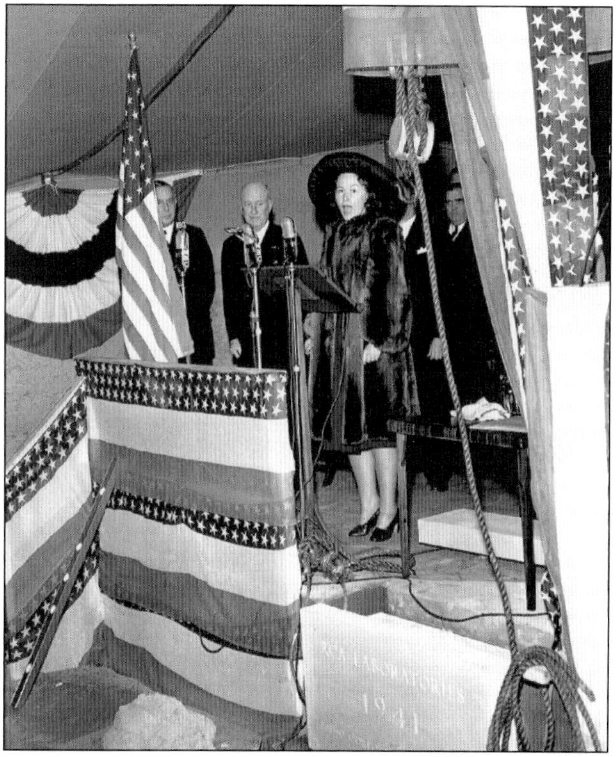

Lucy Monroe sings "The Star-Spangled Banner" into NBC velocity microphones invented by RCA Labs' acoustics engineer Harry Olson. A minor pop singer in the 1930s, Monroe earned a reputation as the finest singer of the national anthem, which she sang more than 10,000 times in her career for presidents, the New York Yankees, and RCA, among others.

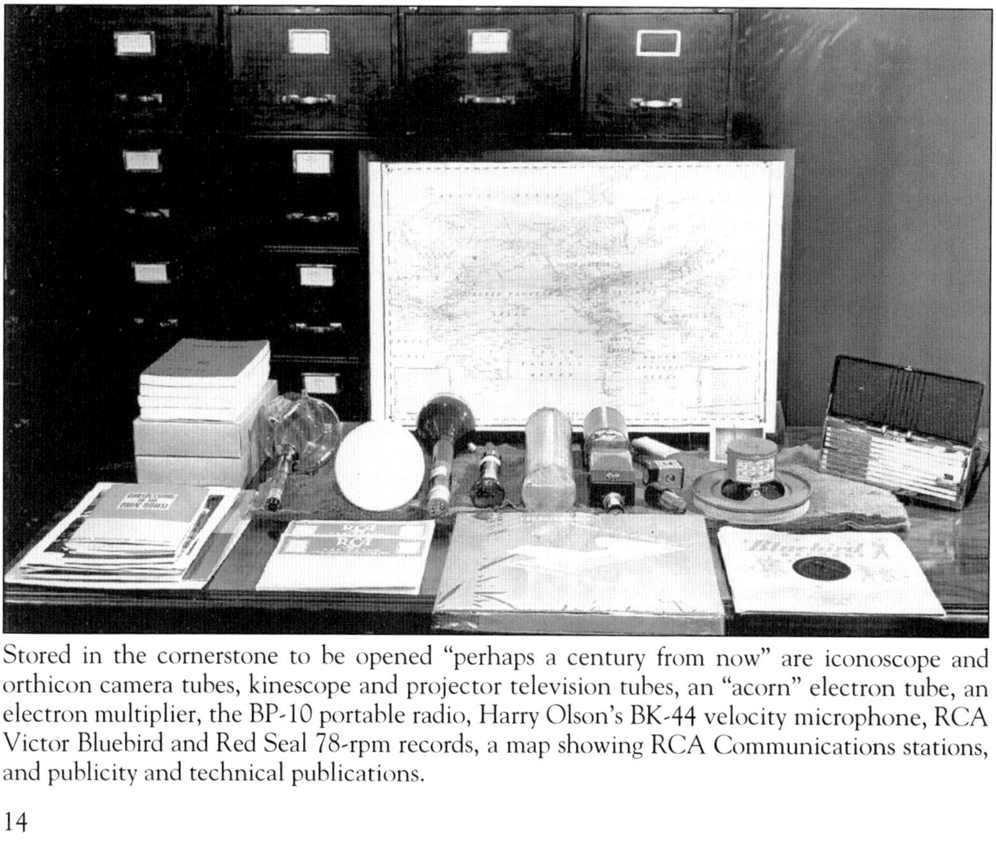

Stored in the cornerstone to be opened "perhaps a century from now" are iconoscope and orthicon camera tubes, kinescope and projector television tubes, an "acorn" electron tube, an electron multiplier, the BP-10 portable radio, Harry Olson's BK-44 velocity microphone, RCA Victor Bluebird and Red Seal 78-rpm records, a map showing RCA Communications stations, and publicity and technical publications.

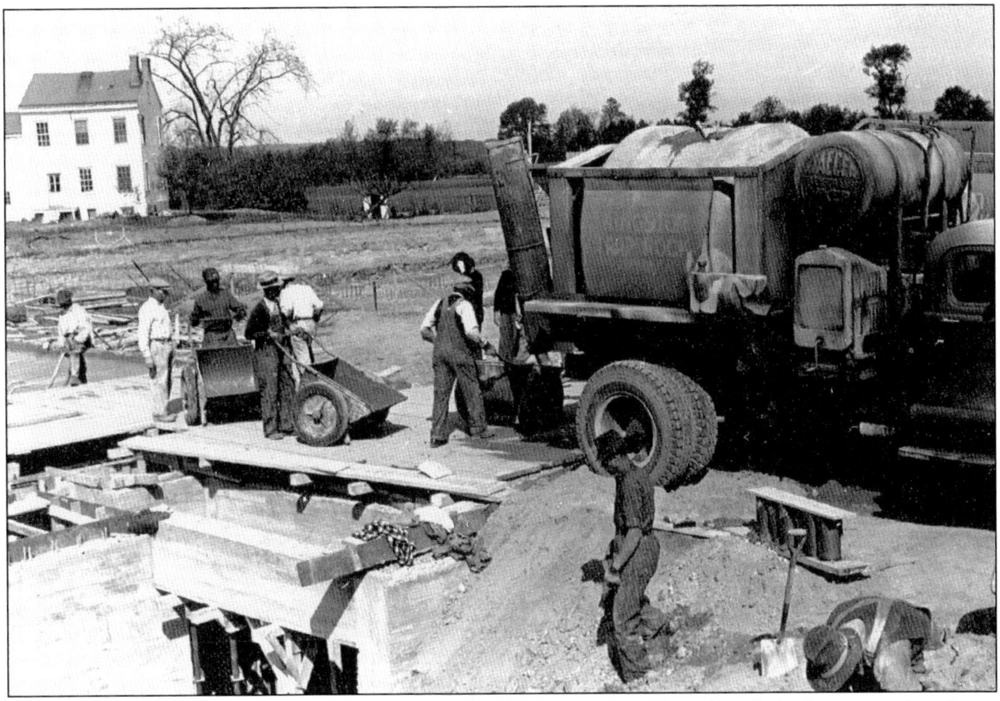

Concrete for the labs' basement was provided by Kingston Trap Rock Industries, already in business for 80 years at the time of this photograph.

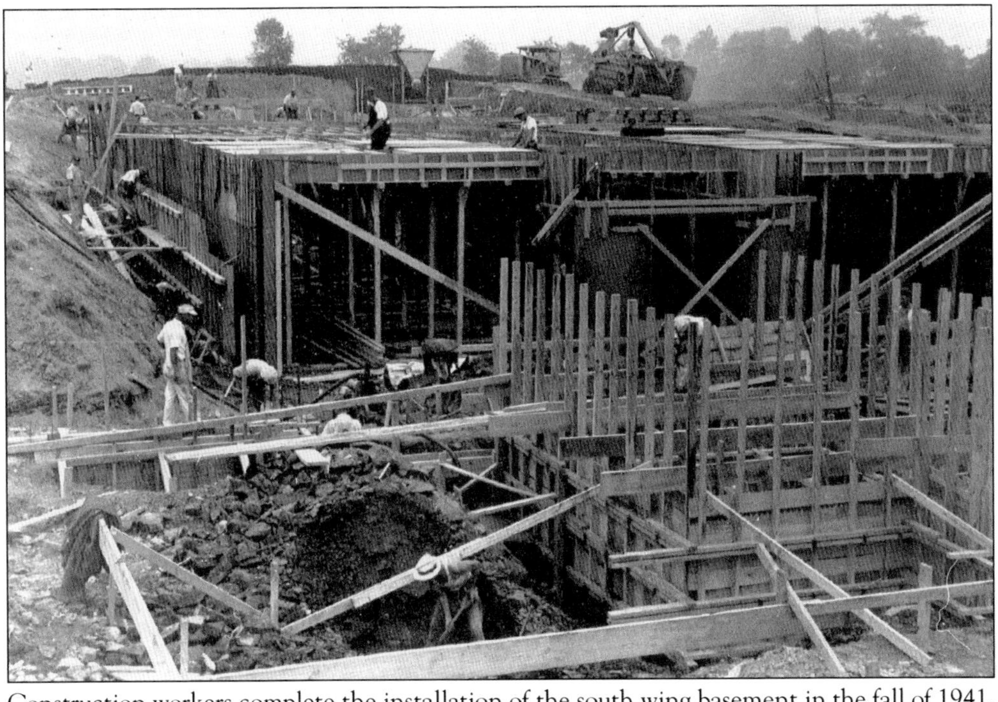

Construction workers complete the installation of the south wing basement in the fall of 1941.

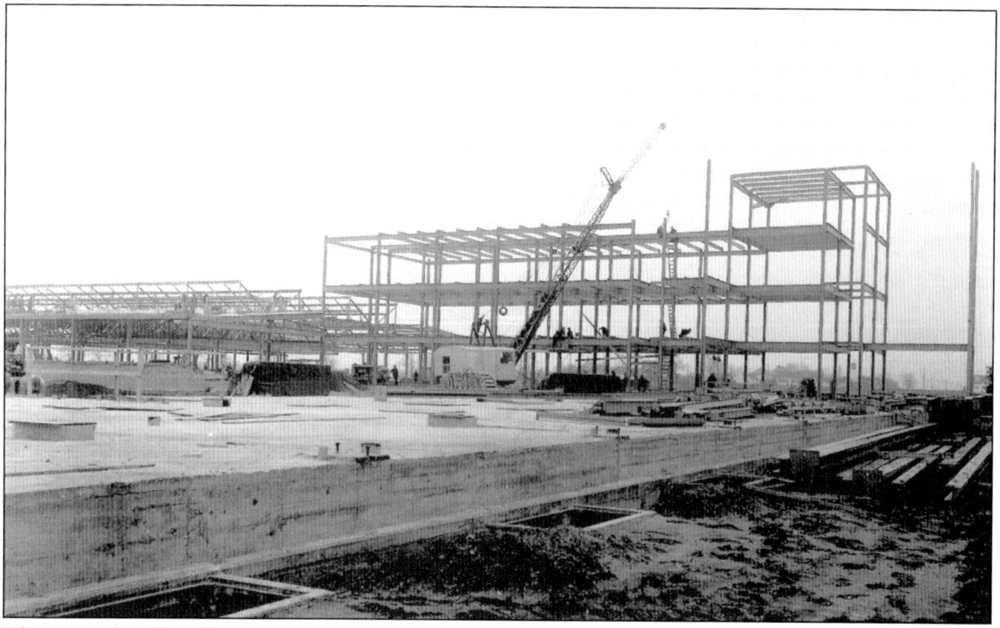

Three weeks after the cornerstone ceremony, Japan attacked Pearl Harbor. By mid-December 1941, H.K. Ferguson had erected the frame of the model shop, south wing, and center of the labs.

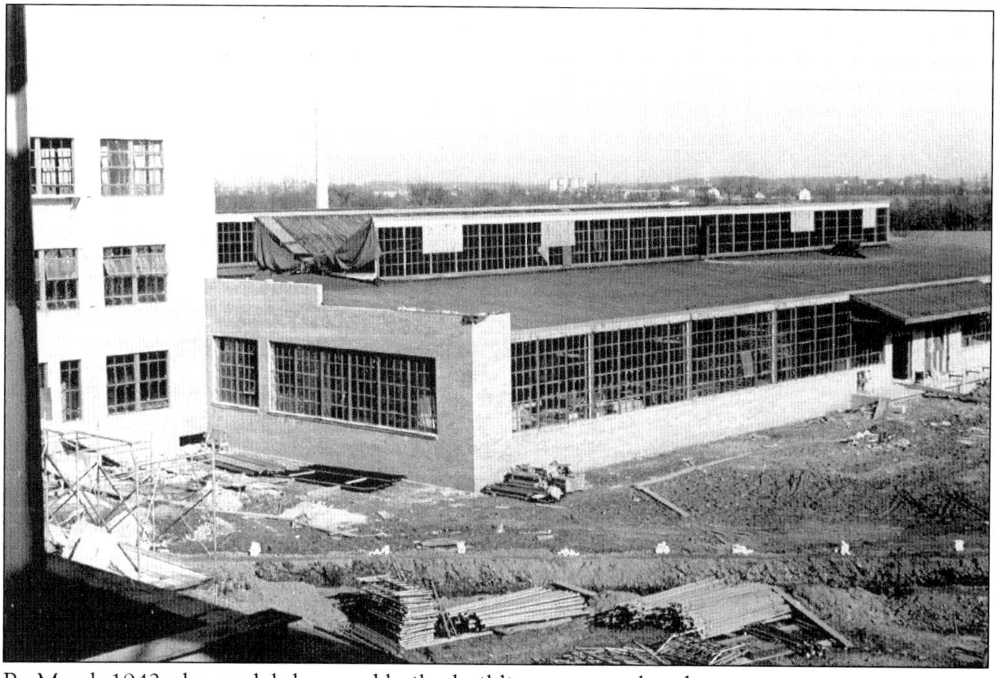

By March 1942, the model shop and boiler building were enclosed.

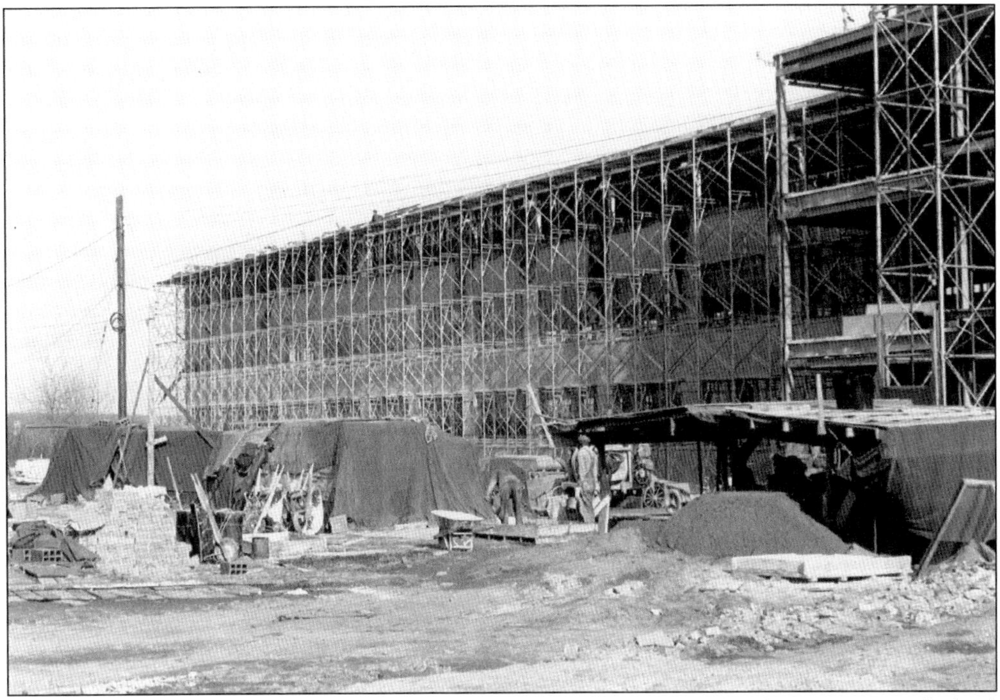

At the same time, bricklaying continued on the east wing (above). The west wing (below) had stairs.

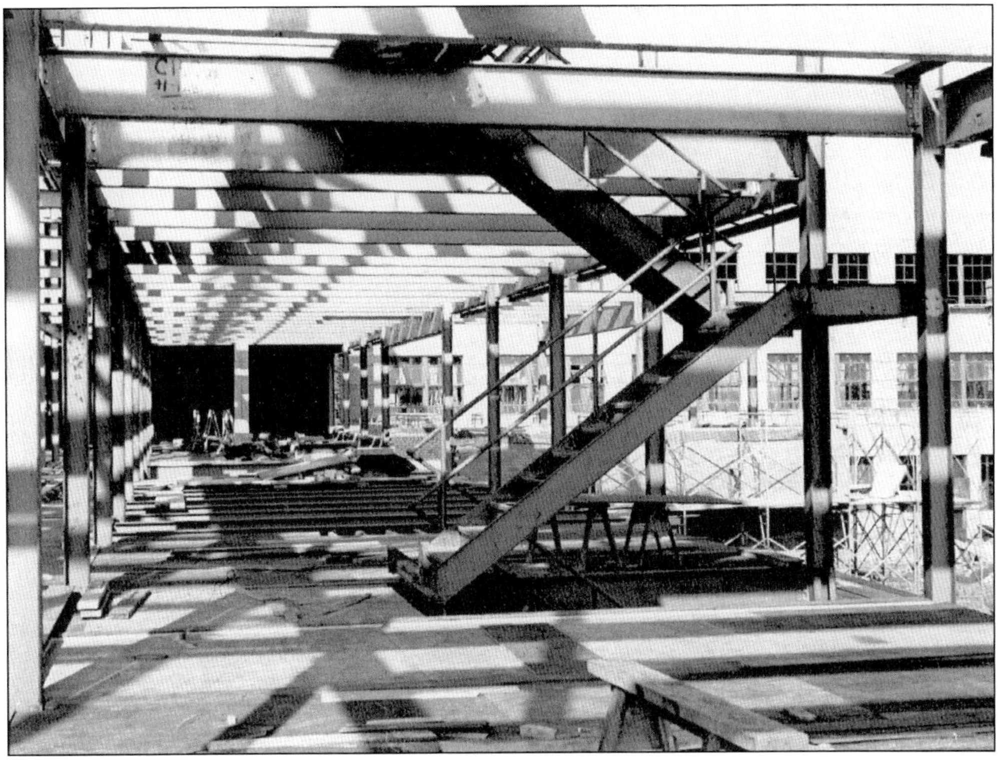

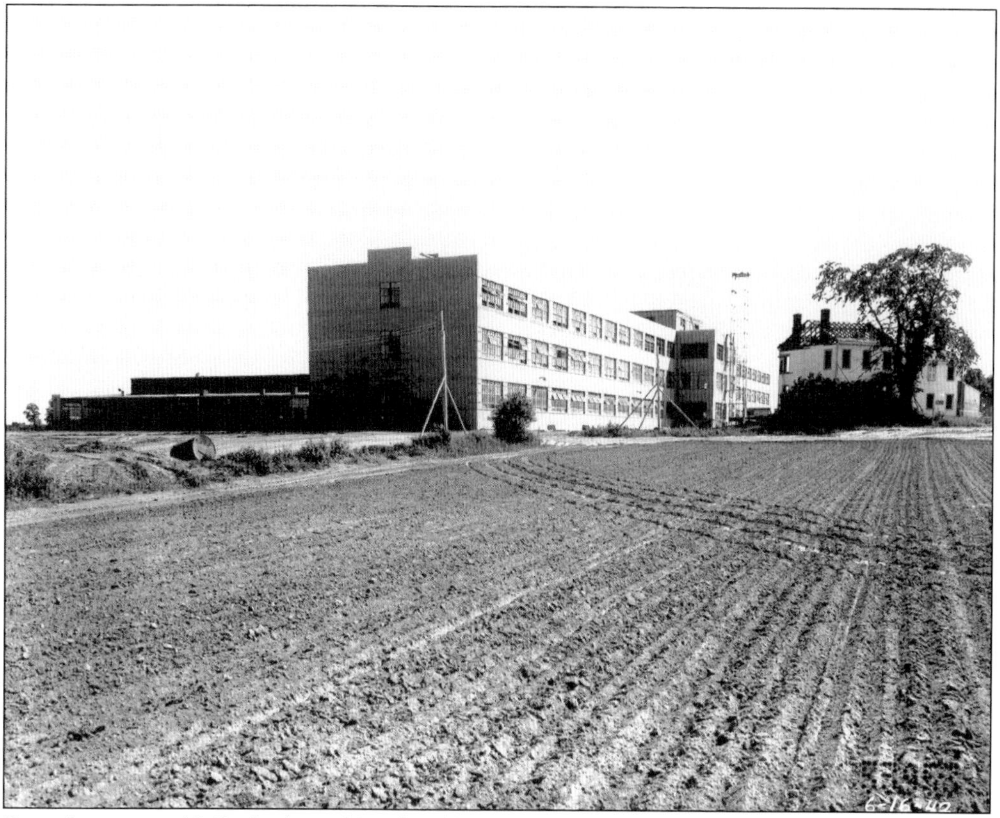

By early summer 1942, the last of Sarah Olden's tenants had tilled his sublet property one final time, and RCA scientists and engineers began moving into their new building.

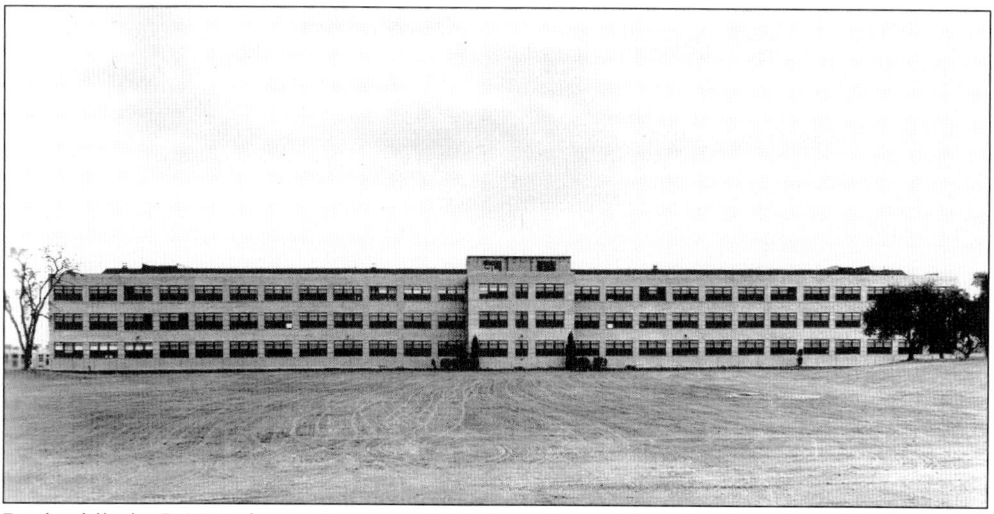

By the fall, the RCA Laboratories were visible from Route 1 but closed to the public.

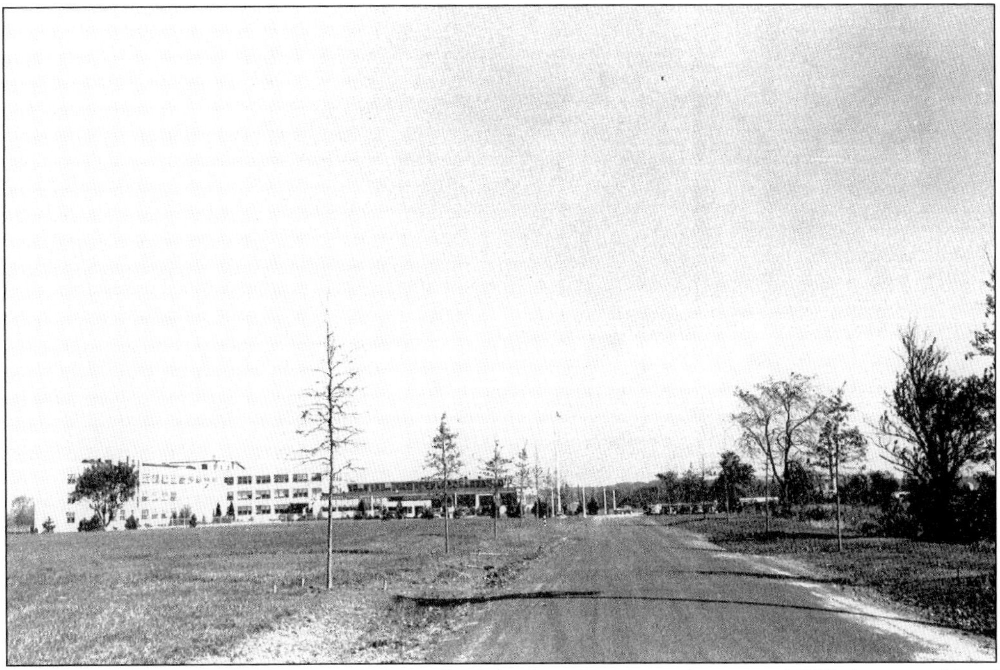

This photograph shows the approach to the labs from Fairview Avenue off Washington Road in mid-1942.

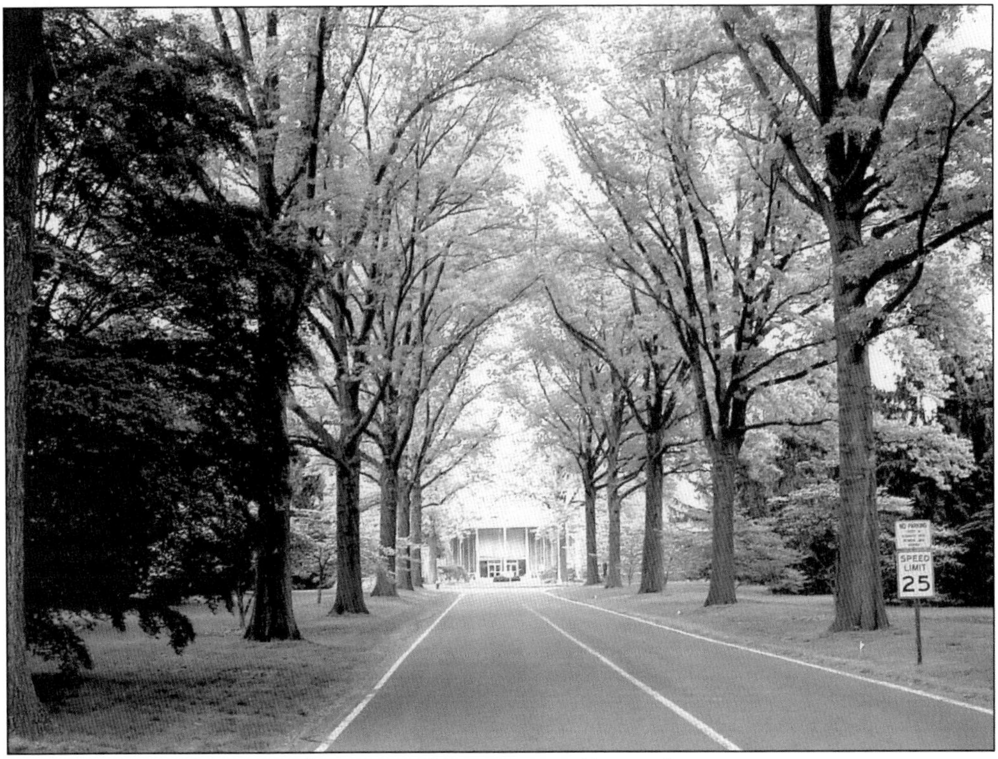

This view shows the approach to Sarnoff Corporation 60 years later.

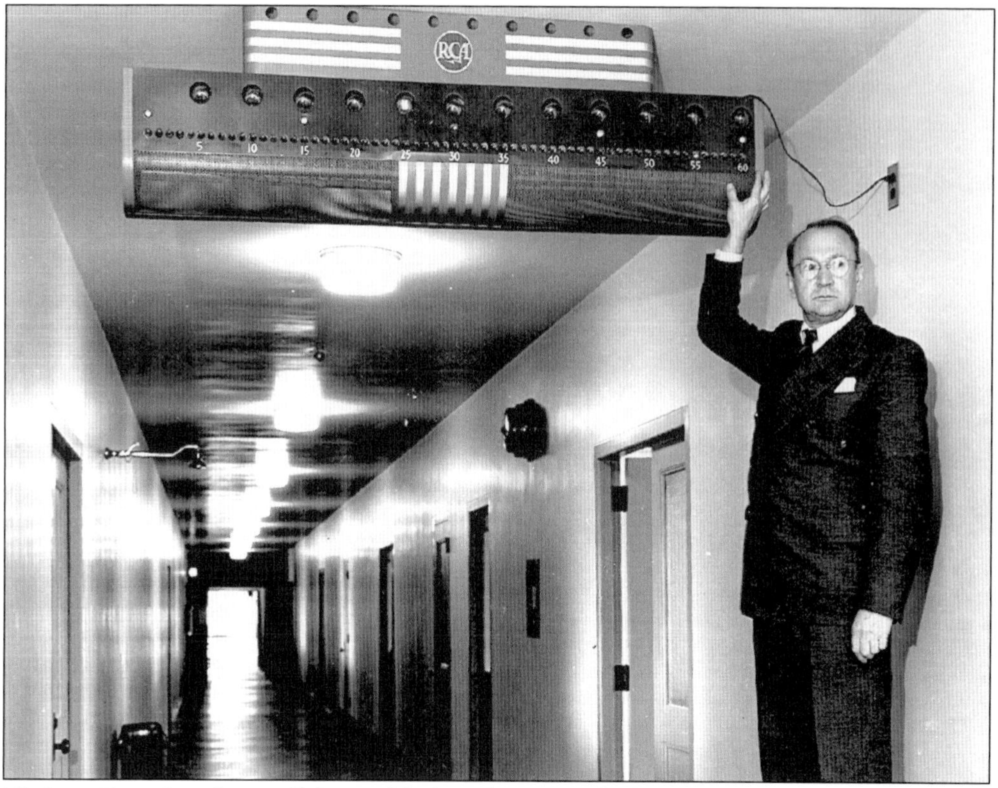

Vladimir Zworykin shows off the world's first electronic clock, installed on the first floor of the labs. Based on patents conceived by Les Flory that helped point the way to digital electronic computing, the clock had no moving parts. One hundred seventy electron tubes counted and stored the pulses from standard 60-hertz electric current to track hours, minutes, and seconds. A false switch was added later to keep Zworykin from tinkering with its operation.

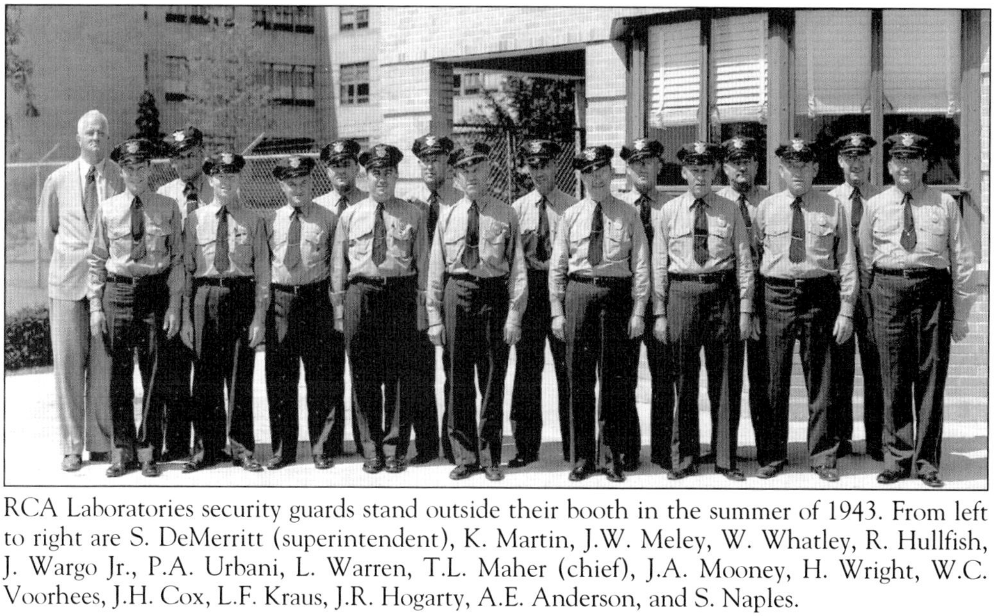

RCA Laboratories security guards stand outside their booth in the summer of 1943. From left to right are S. DeMerritt (superintendent), K. Martin, J.W. Meley, W. Whatley, R. Hullfish, J. Wargo Jr., P.A. Urbani, L. Warren, T.L. Maher (chief), J.A. Mooney, H. Wright, W.C. Voorhees, J.H. Cox, L.F. Kraus, J.R. Hogarty, A.E. Anderson, and S. Naples.

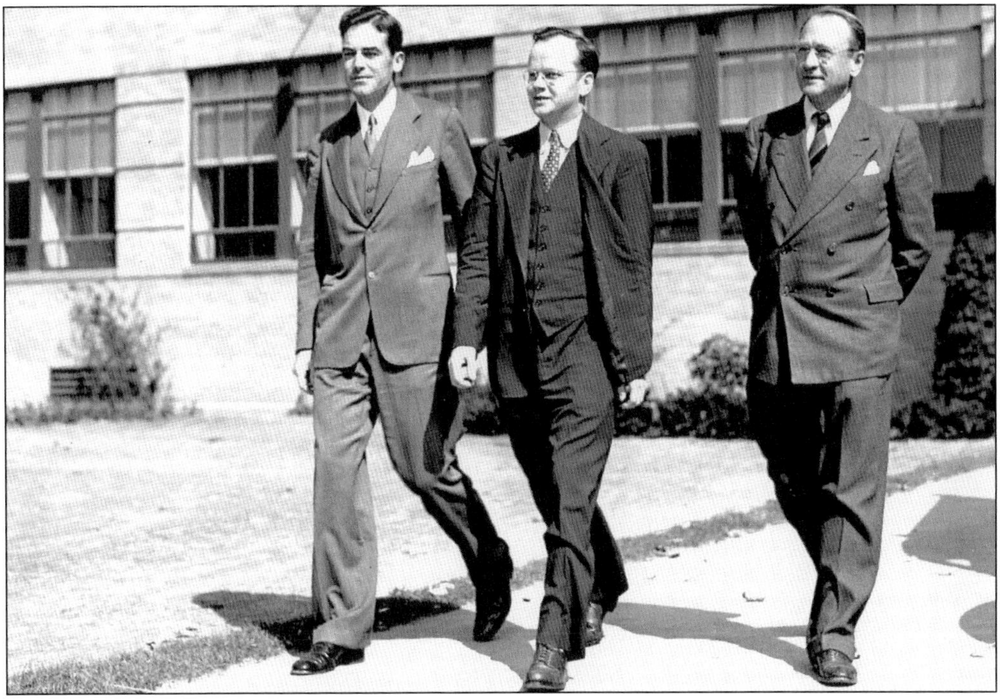

RCA Laboratories director Elmer Engstrom, flanked by associate director Browder J. Thompson (left) and Vladimir K. Zworykin, exits the front door of the labs in the summer of 1942. Previously, Engstrom administered RCA Labs from the Victor Division in Camden; Zworykin had directed Electronic Systems R&D in Camden while Thompson oversaw electron tube research in Harrison.

Browder J. Thompson, a rising star in the RCA hierarchy, was credited with leading the innovation of ever-higher frequency tubes in the second half of the 1930s. Thompson died in Italy in 1944 while researching the realities of air-to-ground radar usage when German antiaircraft gunners shot down the plane in which he was flying.

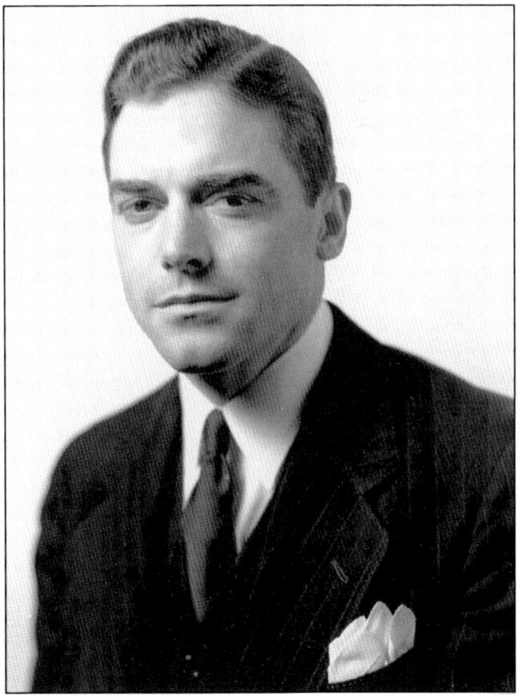

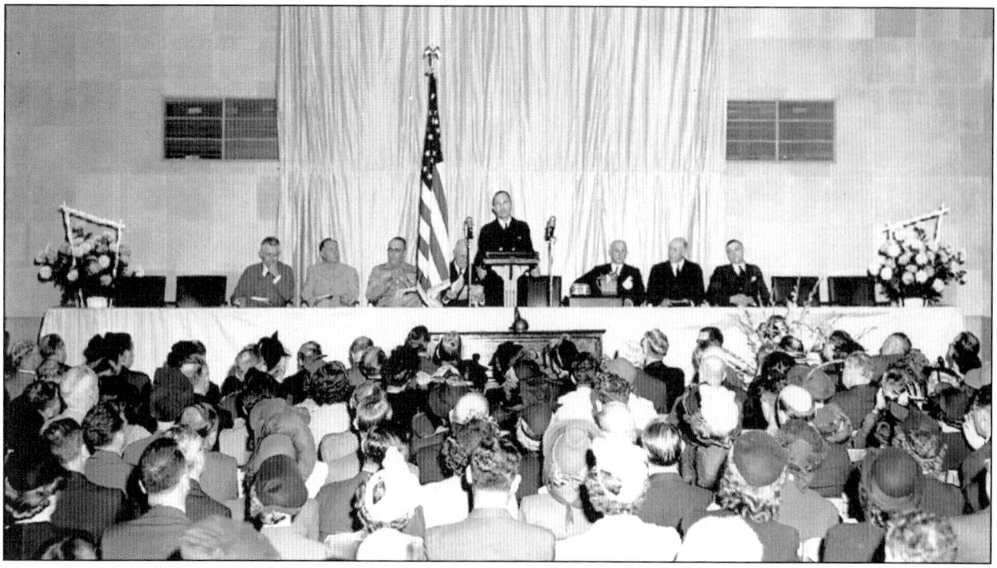

Comdr. A.M. Granum of the U.S. Navy speaks at the dedication of the laboratories on September 27, 1942. From left to right behind him are Princeton University president Harold Dodds, RCA president David Sarnoff, Maj. Gen. Dawson Olmstead of the U.S. Army Signal Corps, RCA chairman James G. Harbord, RCA vice president Otto Schairer, Institute of Advanced Study director Frank Aydelotte, and chief engineer E.K. Jett of the Federal Communications Commission.

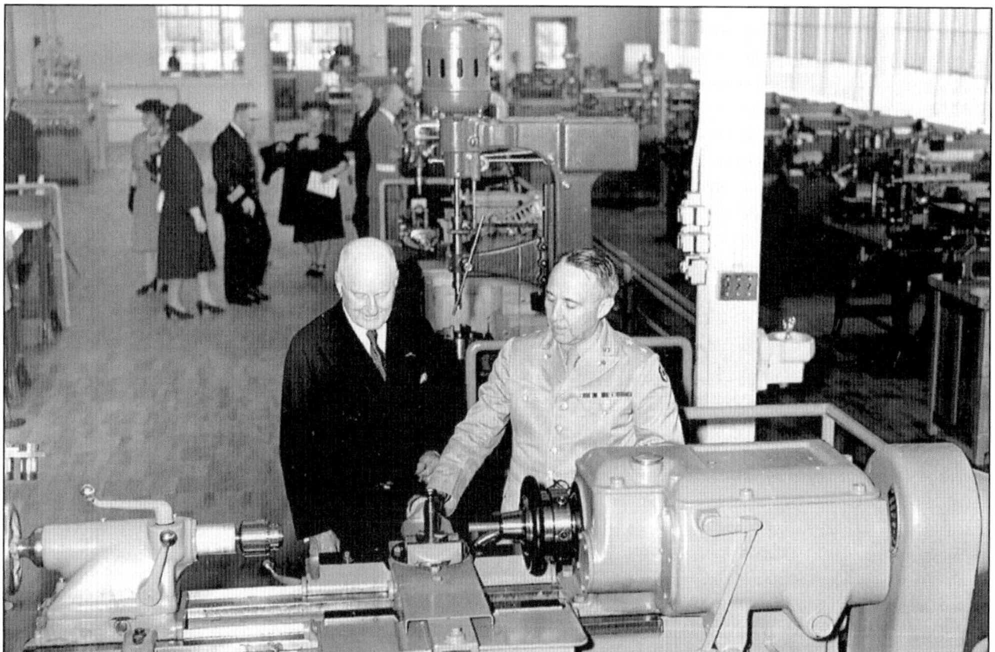

After the ceremony in the television research studio in the east wing, Gen. James G. Harbord (left) and Maj. Gen. Dawson Olmstead, chief signal officer of the Signal Corps, admire one of the model shop's new lathes. In the background are some of the last civilians to set foot in the laboratories for the duration of the war.

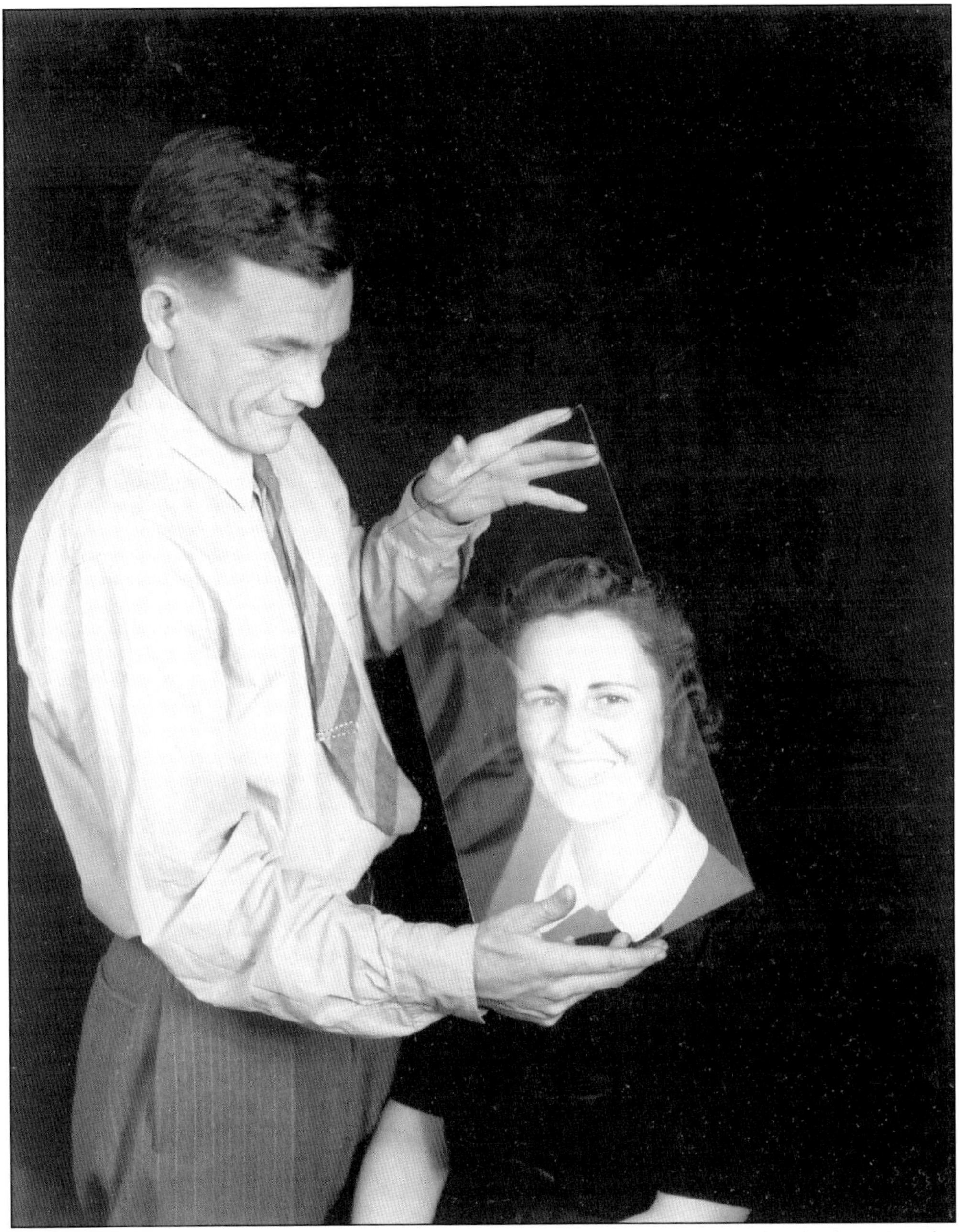

Ted Nicol shows off the nonreflective glass he developed for projected radar displays. For the surface studies necessary to make the glass, he drew on RCA's new electron microscope, which James Hillier continued to improve and elaborate on during the war.

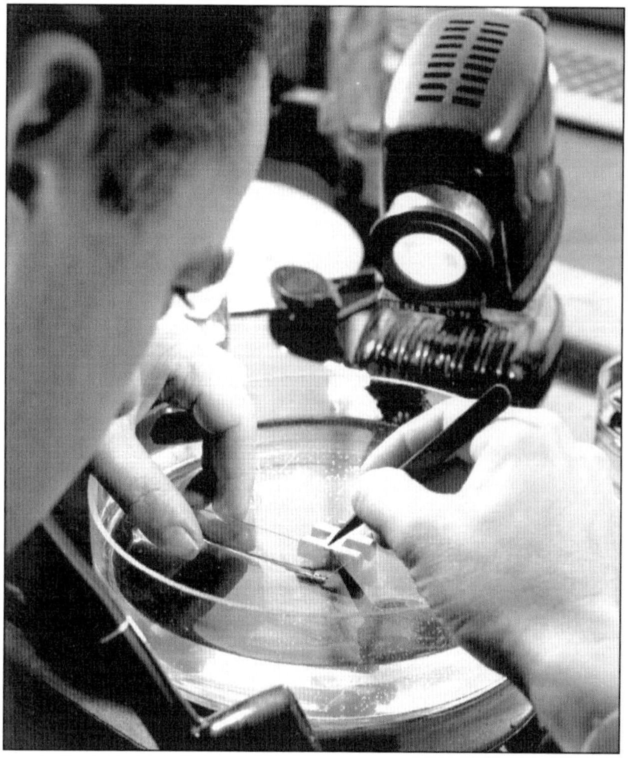

James Hillier prepares a specimen in 1945. Beyond his efforts to improve and extend the capabilities of electron microscopy, Hillier also acted as a bridge to biologists, physiologists, and medical doctors. A physicist by training, he promoted the benefits of the new tool and learned for himself some of the challenges of attaining useful images from the mix of organic material and 50-kilovolt electrons.

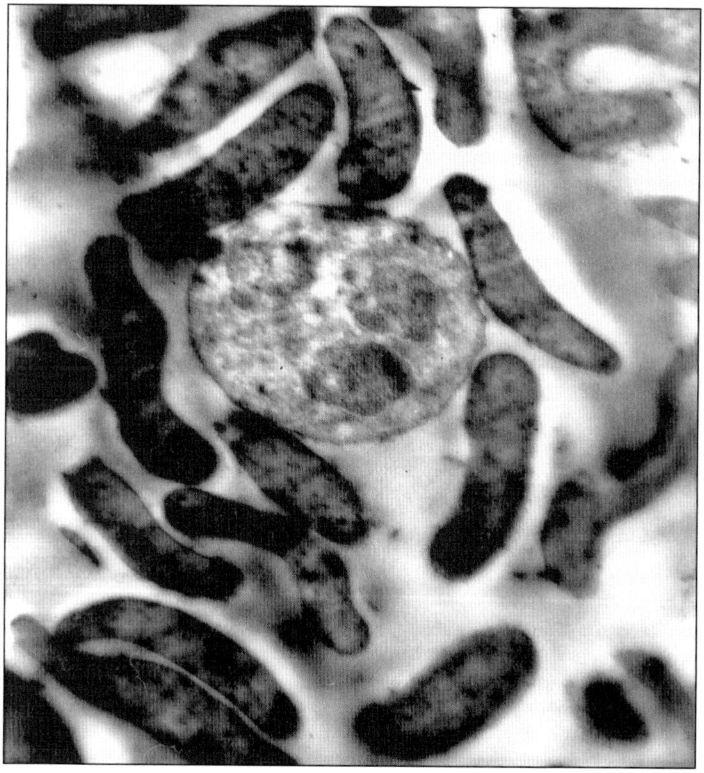

Magnified 10,000 times, a white blood cell is surrounded by red blood platelets.

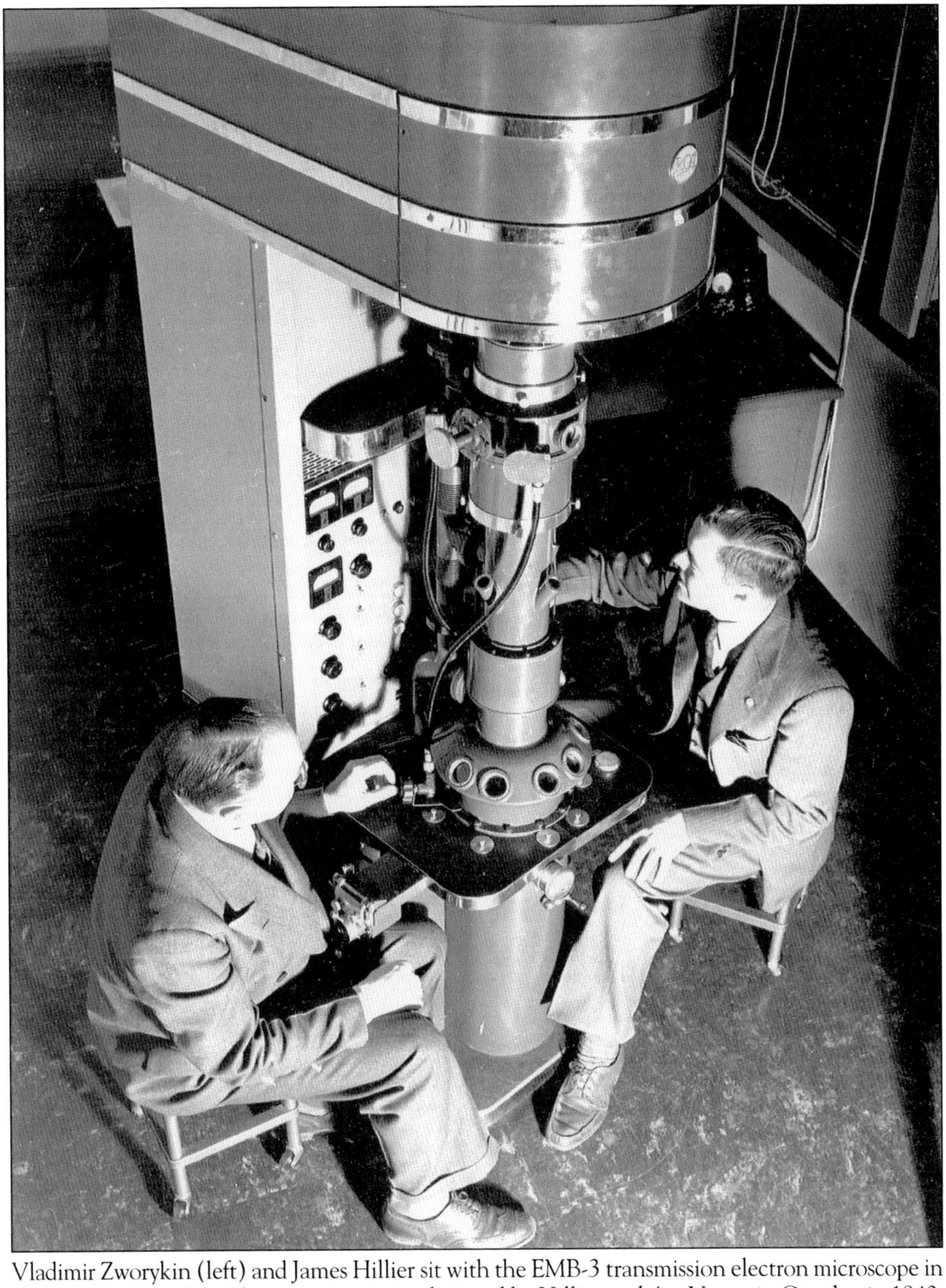

Vladimir Zworykin (left) and James Hillier sit with the EMB-3 transmission electron microscope in Hillier's east wing lab. The microscope was designed by Hillier and Art Vance in Camden in 1940. RCA Victor manufactured and sold 56 of them for use initially in viral and particle research.

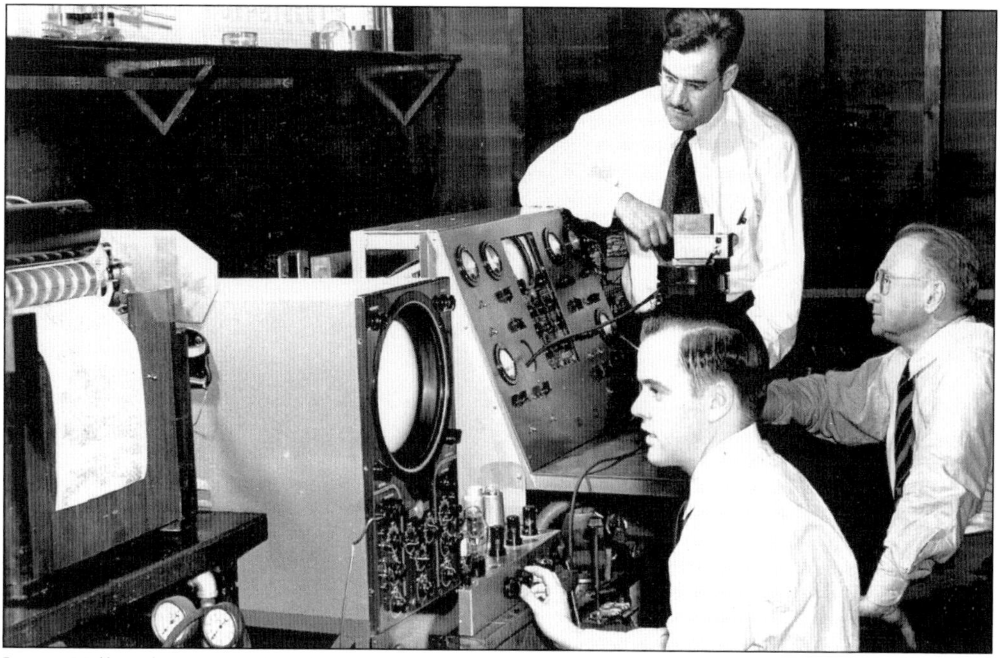

James Hillier (sitting in front), Richard L. Snyder (standing), and Vladimir Zworykin operate the world's first scanning electron microscope designed to observe surfaces. They drew on their own knowledge of electron optics and television technologies, and also on innovations in facsimile printing led by Maurice Artzt of the labs. The relative complexity of the system and lack of demand for high-contrast surfaces limited commercialization until 1967.

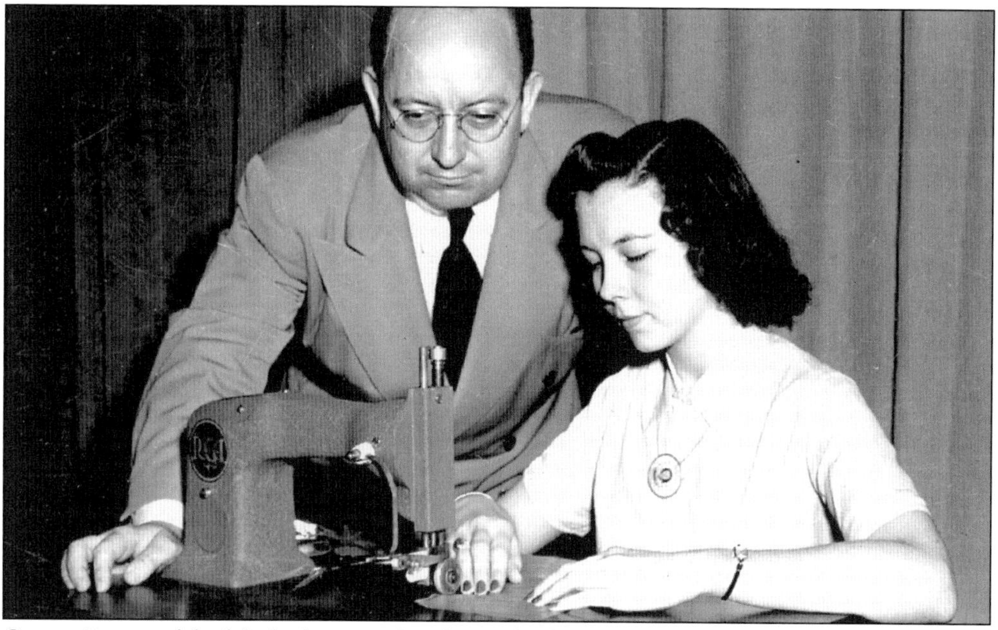

George Brown of the labs looks on as Carol Weatherly of the RCA Victor Division in Camden "sews" a plastic seam with radio-frequency heat in 1943. Brown, Cyril Hoyler, and Rudy Bierwirth pioneered this industrial technique in response to a variety of military needs, in this case sealing sheets of the plastics mass produced for war effort.

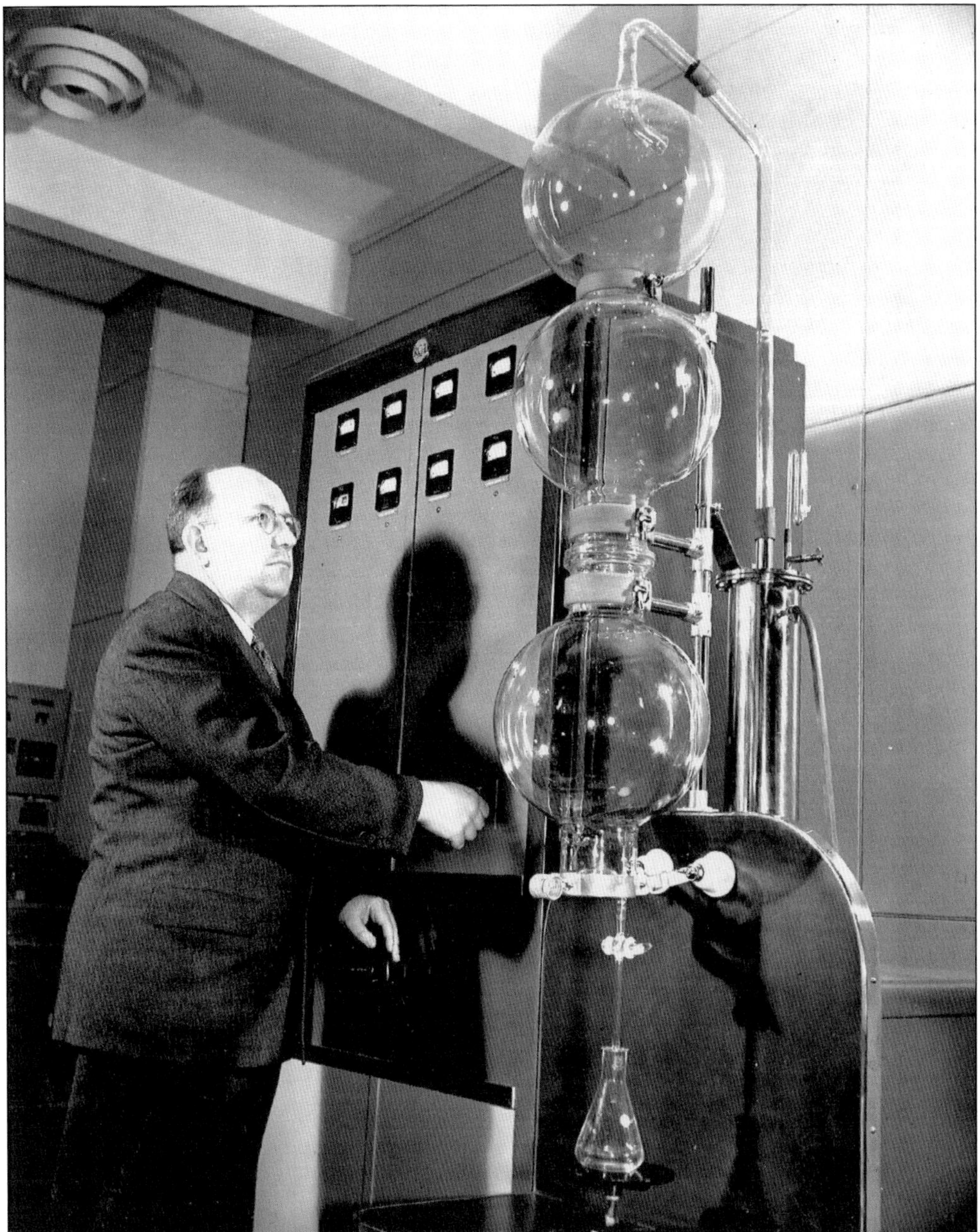

George Brown inspects the machine for the radio-frequency (RF) bulk dehydration of penicillin before the labs ship it up Route 1 to E.R. Squibb & Sons in New Brunswick in July 1944. Moved by the pharmaceutical industry's difficulties in drying out penicillin mold without destroying its potency, Brown, Rudy Bierwirth, and Cyril Hoyler applied their experiences with drying out cakes of rayon by radio-frequency heating.

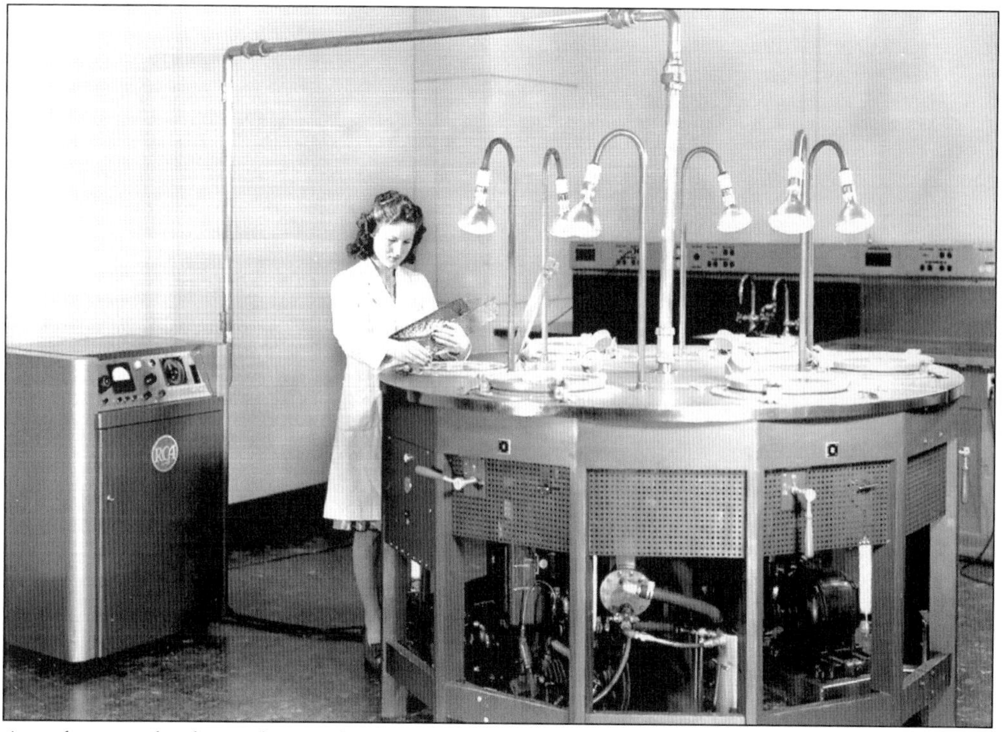

A technician loads up the machine for final dehydration of penicillin. The machine could process 2,040 bottles an hour in January 1945. Shortly afterward, Brown and Wendell Morrison applied for a patent to manufacture penicillin using butyl alcohol in water (U.S. No. 2,536,676), which was widely adopted and rendered the radio-frequency techniques obsolete.

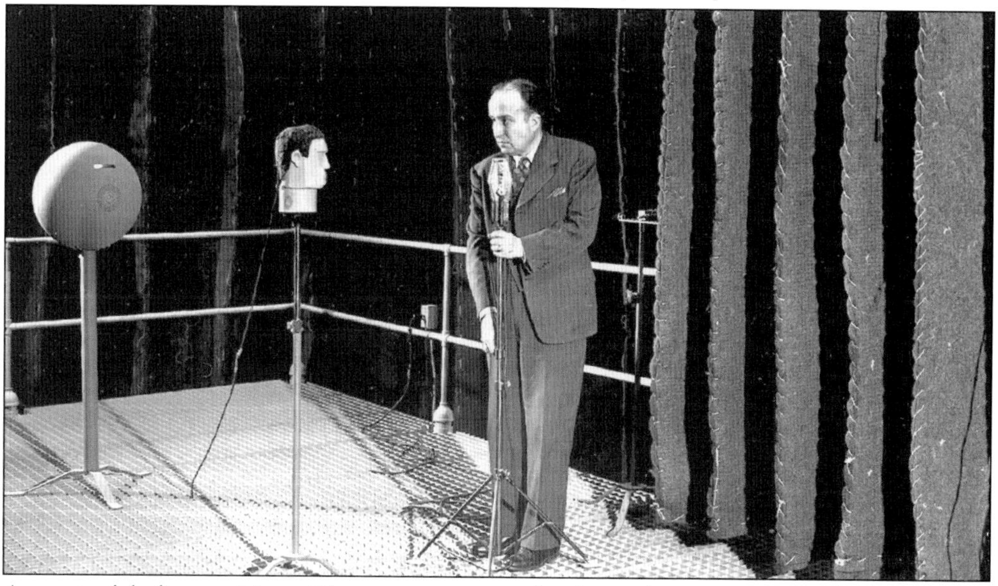

Acoustics lab director Harry Olson sets up an artificial speech experiment in the west wing's three-story anechoic chamber late in 1944. The sphere is another sound source in a room so quiet one could hear one's heart beat when the door closed. Olson took pride in having a free-field room larger than that at Bell Laboratories.

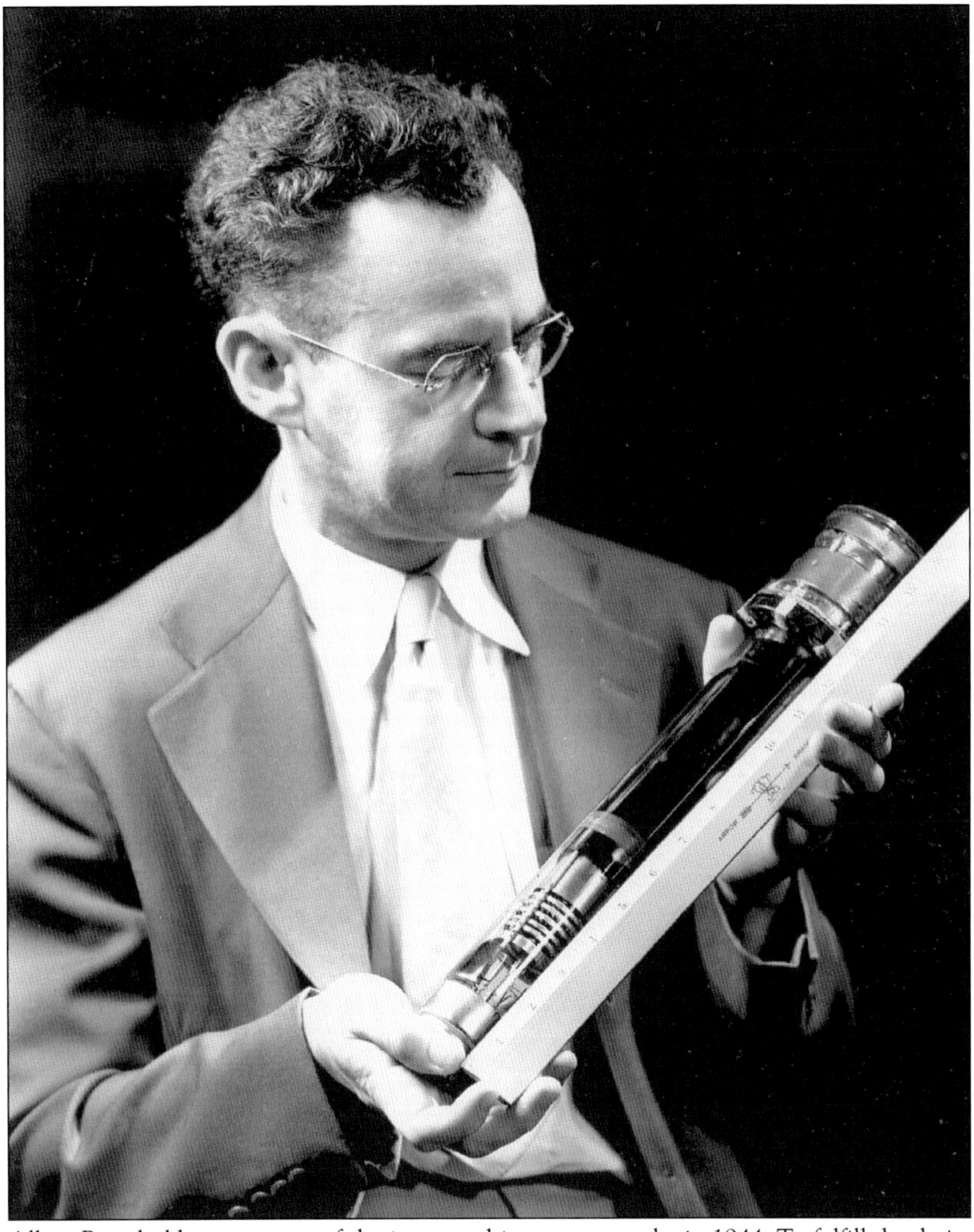

Albert Rose holds a prototype of the image orthicon camera tube in 1944. To fulfill the desire for remotely guided missiles, the military needed a camera able to register distinct images in a wide and changing array of light conditions, over land and sea. The image orthicon offered a light sensitivity 100 to 1,000 times better than that of RCA's iconoscope, the standard prewar television camera tube.

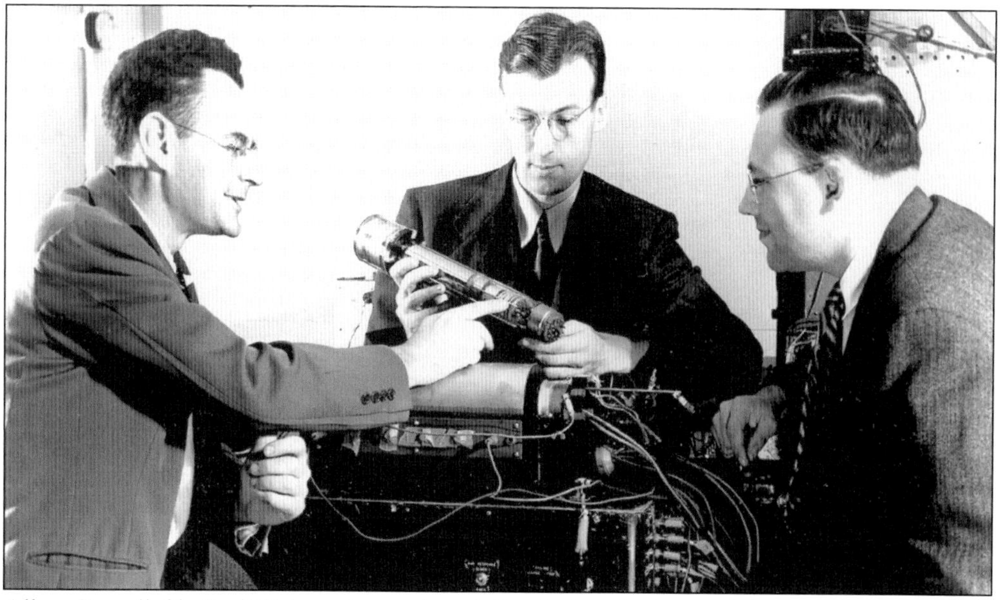

Albert Rose (left), Paul Weimer (center), and Harold Law look over their image orthicon after RCA announced its existence to the public in 1945. The same qualities that made it useful for guided missiles—its low-light sensitivity and extended dynamic range—made it the standard for 20 years in the broadcast television industry, where the "immy" became the namesake for the Emmy in 1948.

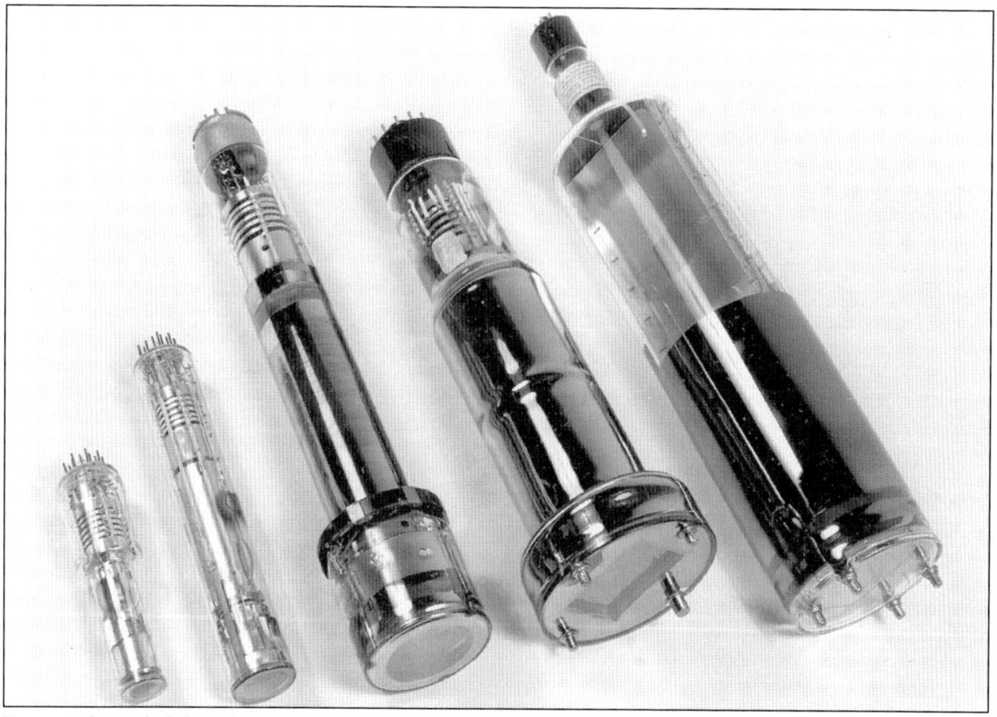

From right to left lies the evolution of video camera tubes at the labs from 1942 to 1945, starting with Rose's orthicon and ending with the experimental MIMO, or miniature image orthicon. Despite the reduction in size, the qualities of imaging from these tubes improved.

These infrared binoculars represent perhaps the world's first head-mounted display, designed by the labs and engineered for a helmet by the Johnson Foundation at the University of Pennsylvania. Staff members at the labs, among them Les Flory, made many of these five-pound displays for night-driving and night-flying and discovered that they worked extremely well at speeds up to 50 miles per hour. Problems arose only in following curves and avoiding local policemen curious about vehicles proceeding with no apparent illumination in the dark.

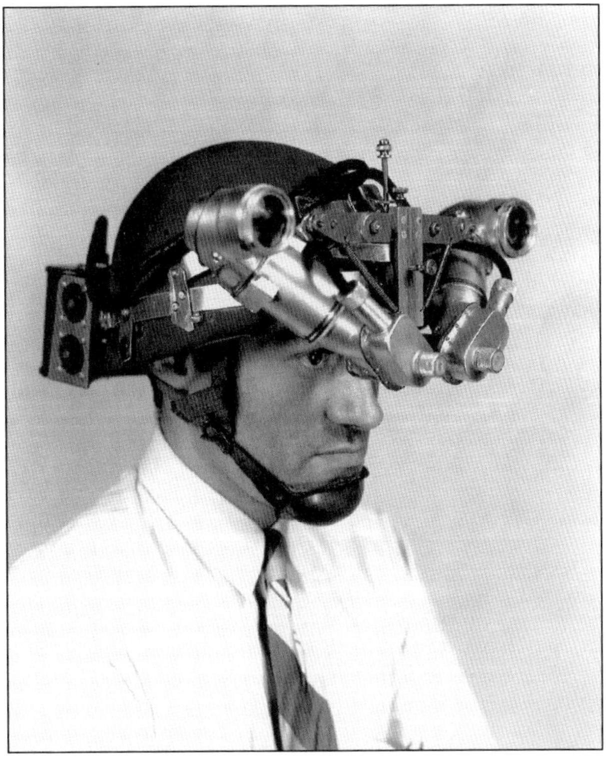

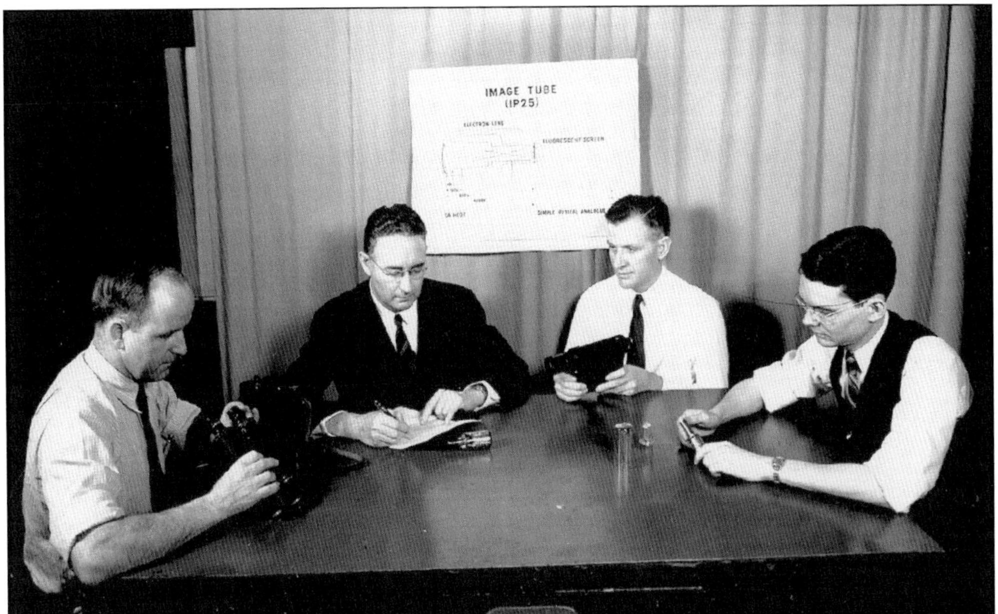

Les Flory (far left), George Morton (second from left), John Ruedy (third from left), and G.L. Krieger show off the infrared image tube they invented and applied to naval and army needs for an electron telescope during the war. Tens of thousands of the tubes were produced for various applications, including infrared signaling with ships miles away, nighttime gunfire, and nighttime driving or flying.

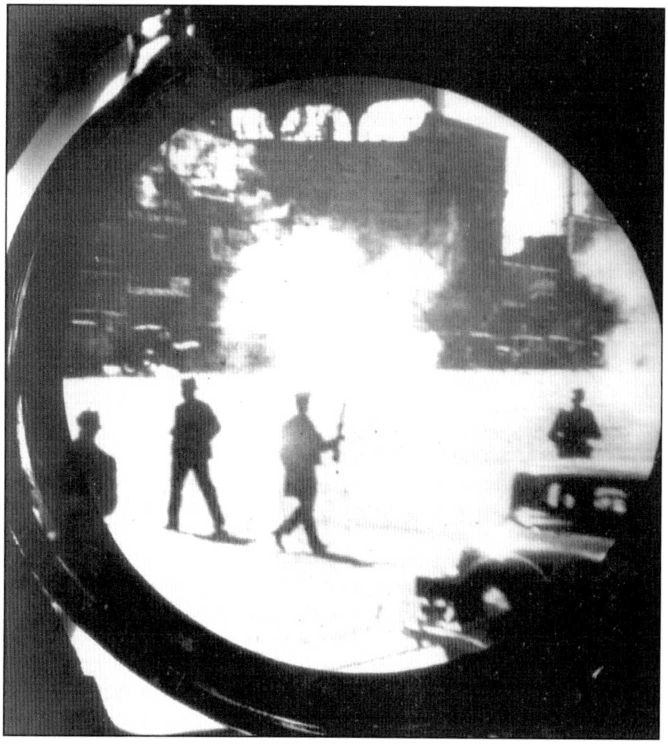

This image shows the view projected on the fluorescent screen in the eyepiece of an electron telescope.

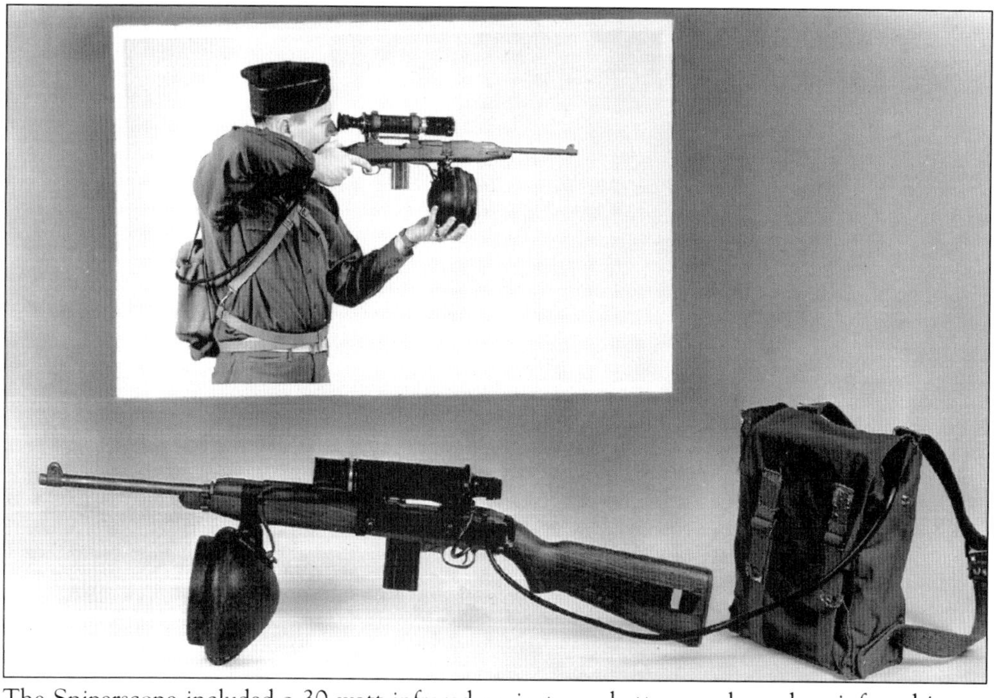

The Sniperscope included a 30-watt infrared projector, a battery pack, and an infrared image tube that served as a gun sight. Fear of losing the technology to the enemy in battle kept it out of service for a year until the invasion of Okinawa in April 1945, where several thousand of the weapons accounted for 30 percent of Japanese casualties in the first months of combat.

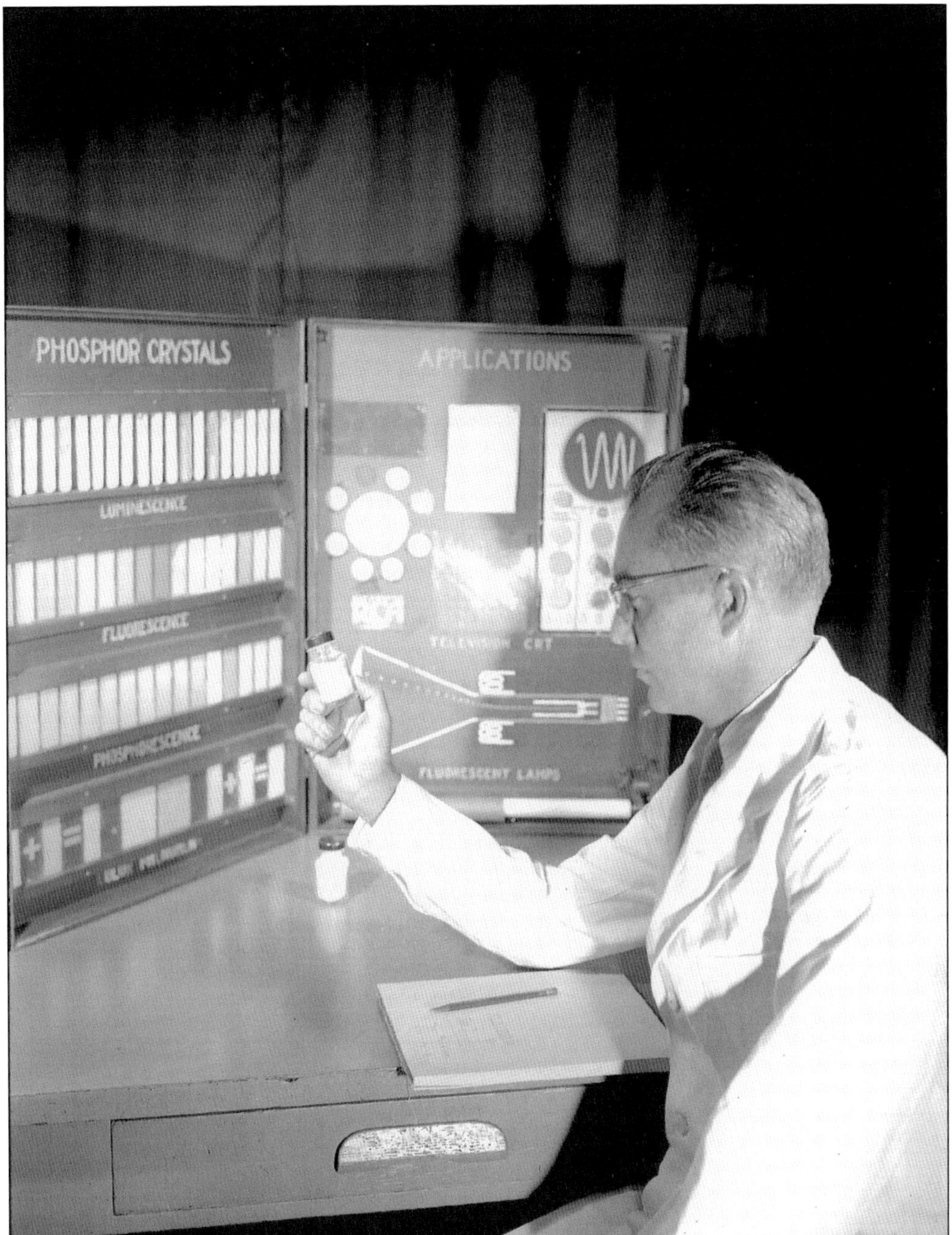

Humboldt W. "Lefty" Leverenz bathes in ultraviolet light with the array of phosphors for which he and the materials laboratory were responsible. Starting in the 1930s, Leverenz developed practical white phosphors for monochrome television—and, coincidentally, fluorescent light bulbs—as well as a variety of long- and short-persistence phosphors for radar displays and the Sniperscope.

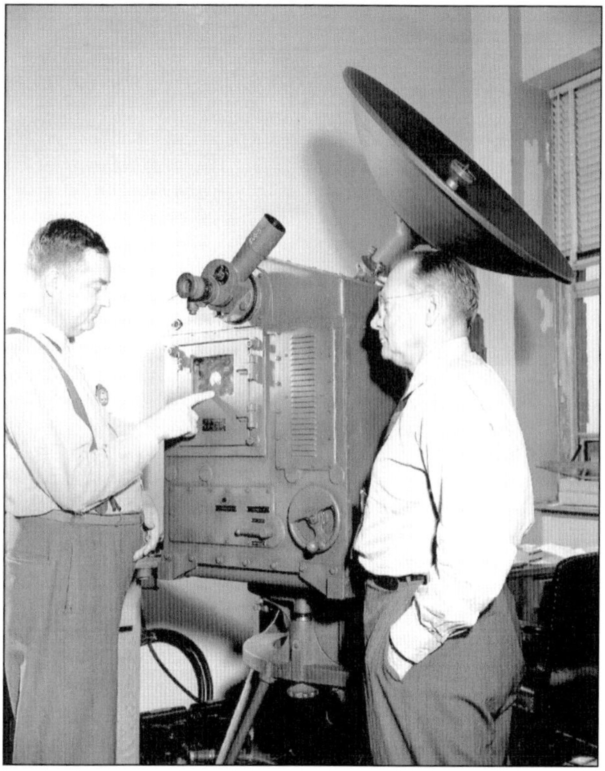

Not everyone can have a radar in his office, but Art Vance does, and he shows off the T35 radar and high-speed director for 40-millimeter antiaircraft guns to Vladimir Zworykin. Vance earned a reputation as perhaps the labs' best circuit designer. He moved from television to the electron microscope to radar plotters and fire control to Project Typhoon after the war.

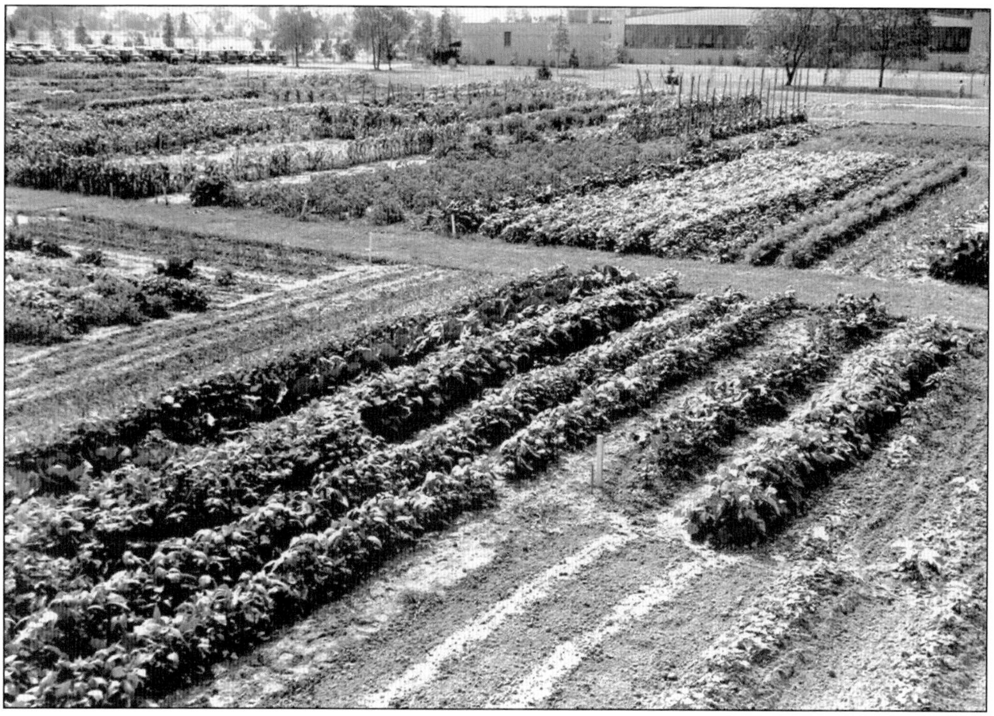

The Victory gardens occupied a considerable part of the labs' backyard, where the parking lot and soccer fields now lie.

Two

To a New World 1946–1954

With the war's end, American businesses and consumers anticipated a new world of peacetime production and high-tech purchases, including the black-and-white televisions much improved by wartime innovations. Not coincidentally, CBS renewed a call for a color television broadcast standard whose home reception was based on a spinning wheel three times the diameter of the picture screen. Between 1947 and 1954, much of the staff invented and innovated the system we use today for capturing, transmitting, receiving, and displaying sound and color video electronically, all within the standard already established for monochrome television.

The staff also pursued innovations in other fields like facsimile, computers, and electronic memory, as well as Gen. David Sarnoff's "birthday presents." RCA Laboratories director Elmer Engstrom joined in the national call for a return to basic research to explore new ways of harnessing the behavior of electrons, those particles—or waves—flashing around atomic nuclei, for practical uses. When the invention of the transistor was announced by Bell Laboratories in northern New Jersey in 1948, RCA Labs semiconductor research embarked on a focused innovation program that resulted in the transistor radio and the solid-state revolution that continues today.

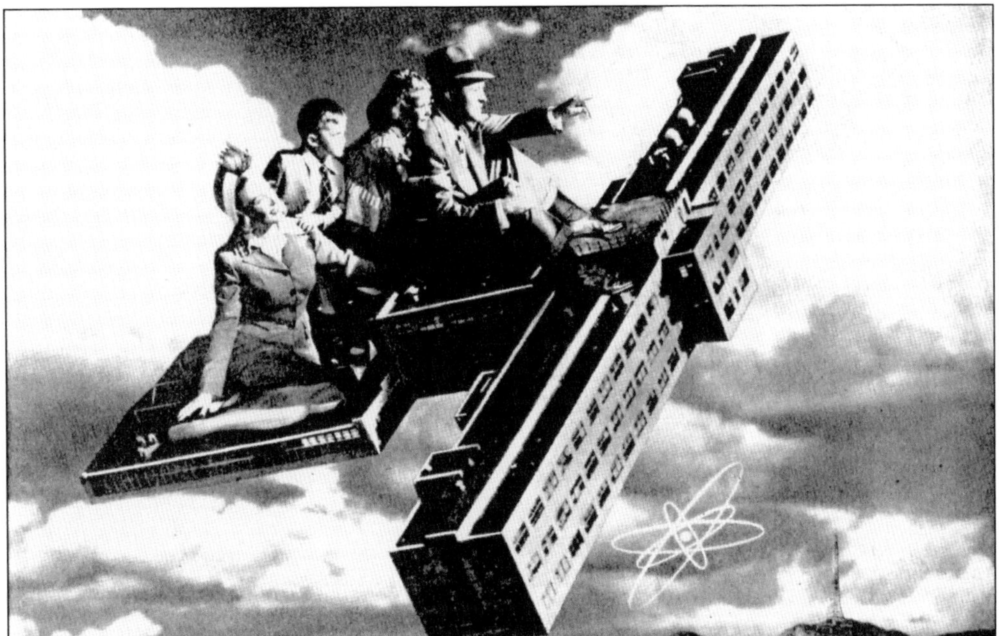

Guided by a microwave relay tower, the mythical American family flies into the future on the roof of the RCA Laboratories. The path is lit by the promise of atomic energy.

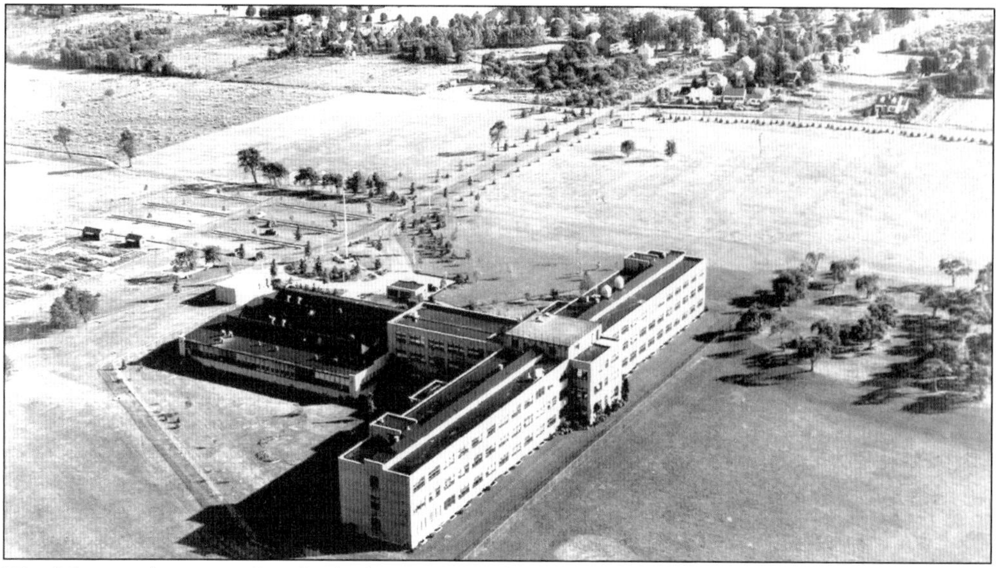

The labs are shown as they looked at war's end. Director Elmer Engstrom was very reluctant to have antennas or radomes break up the lines of the roof, but more pragmatic staff members overcame his reservations.

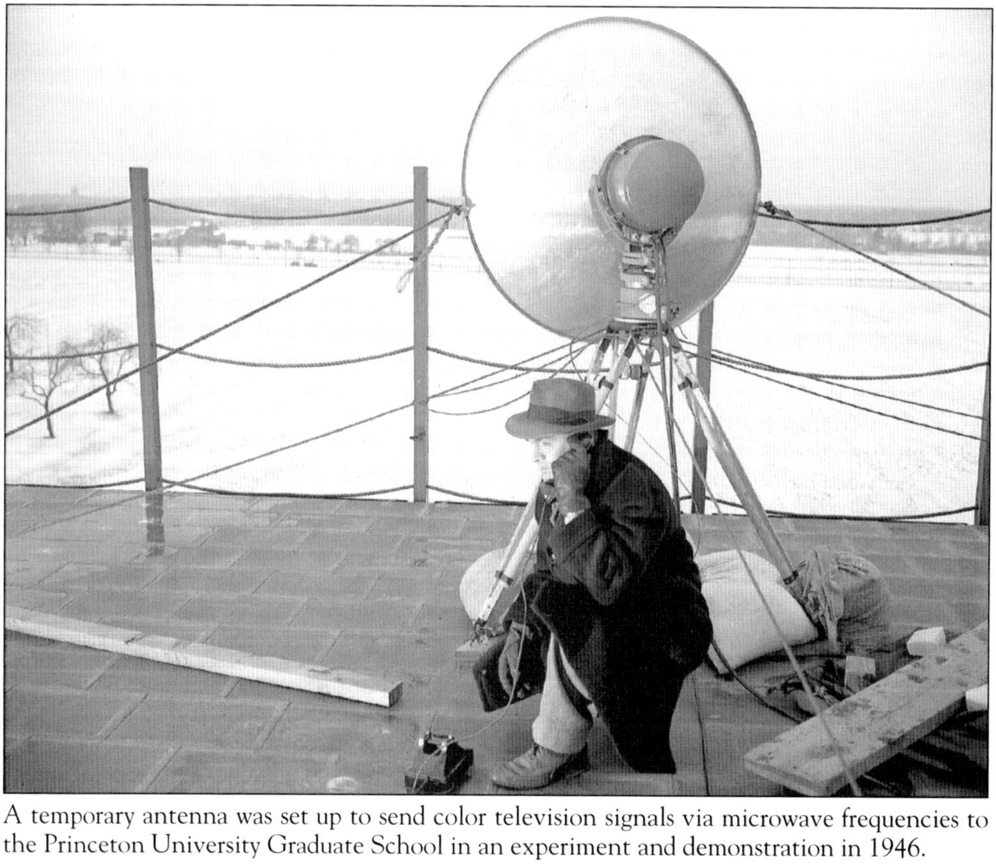

A temporary antenna was set up to send color television signals via microwave frequencies to the Princeton University Graduate School in an experiment and demonstration in 1946.

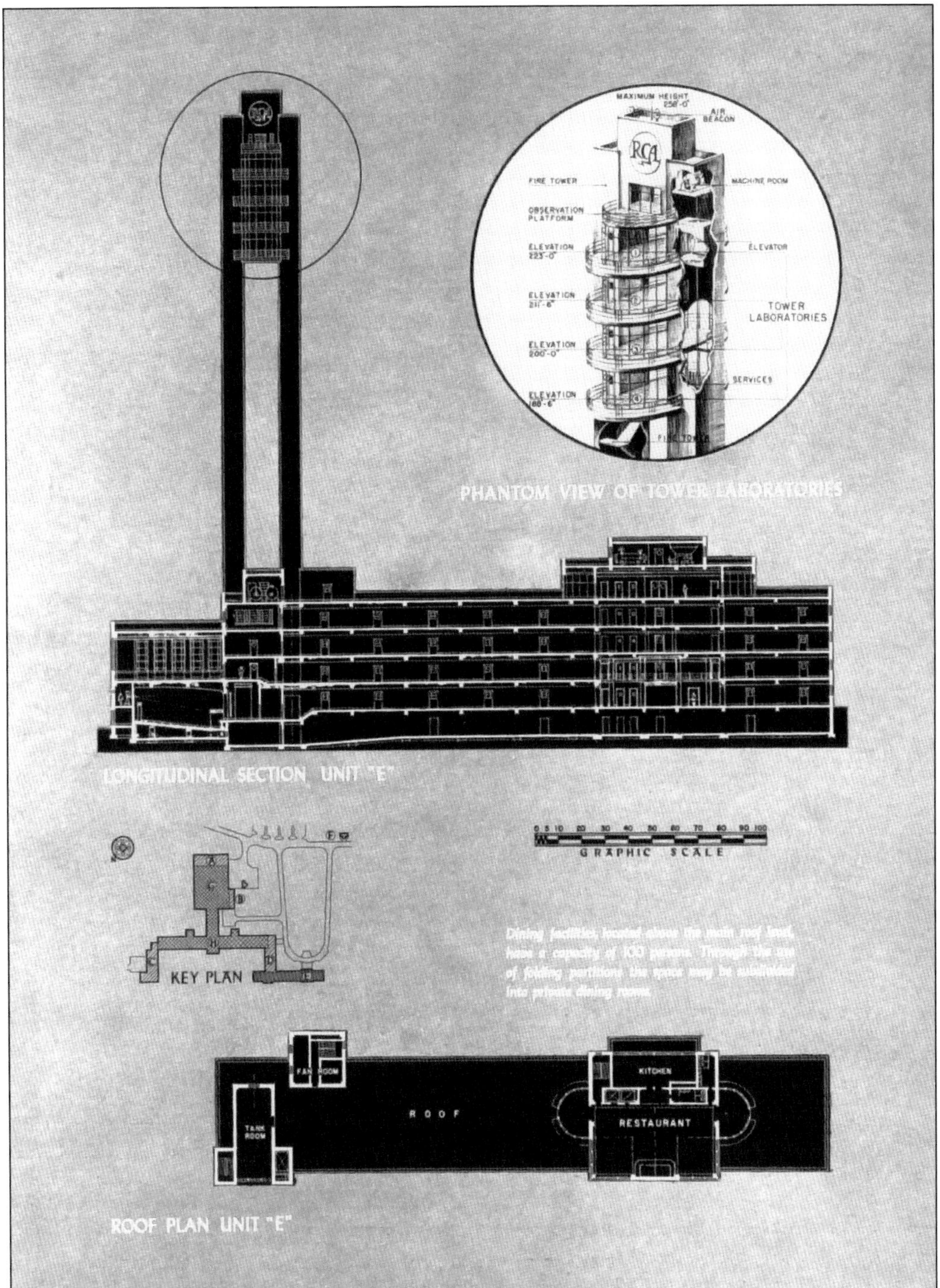

Early in 1946, RCA commissioned Shreve, Harmon and Lamb to develop an architectural plan based on anticipated needs in television and high-frequency electronics research, and for expanded support staff. While some of the report served as the basis for construction through much of the 1950s, the 250-foot tower for antenna and propagation studies was abandoned as cost-inefficient. So too was the rooftop cafeteria on the west wing extension.

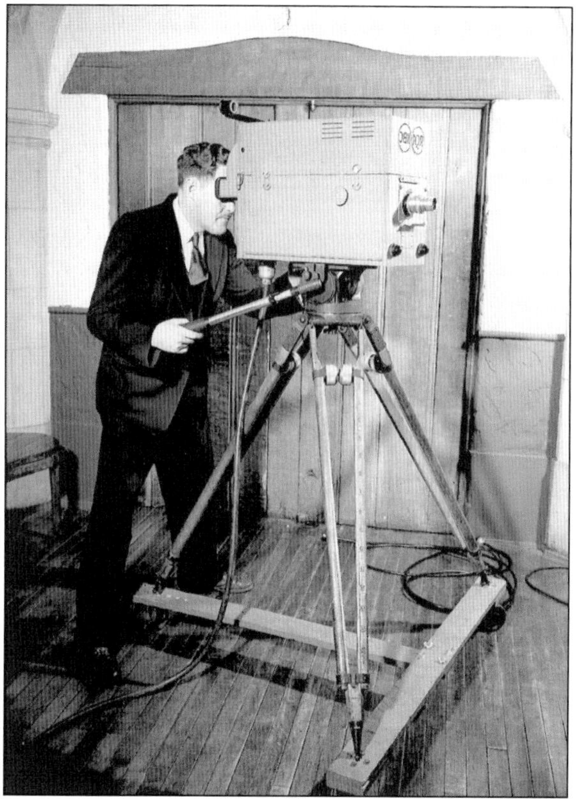

Seen here is Richard Webb's first electronic color television camera, built for tests in Princeton in the fall of 1945.

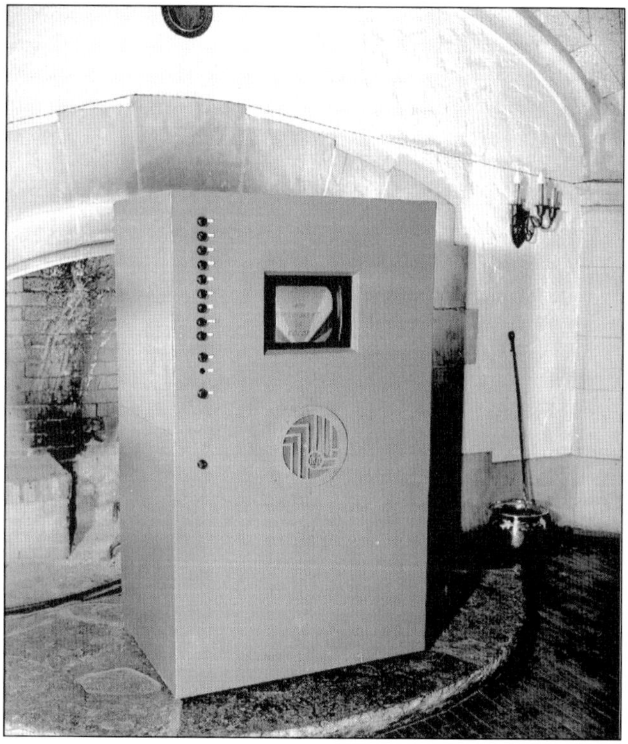

The RCA Labs' field-sequential color television set is ready for its last experiment with an electromechanical system. Behind the screen, a drum of 12 polarized color-filtered mirrors rotates around the display tube. For these final demonstrations at the Princeton University Graduate School, RCA's engineers made 3-D glasses out of sunglasses bought in downtown Princeton.

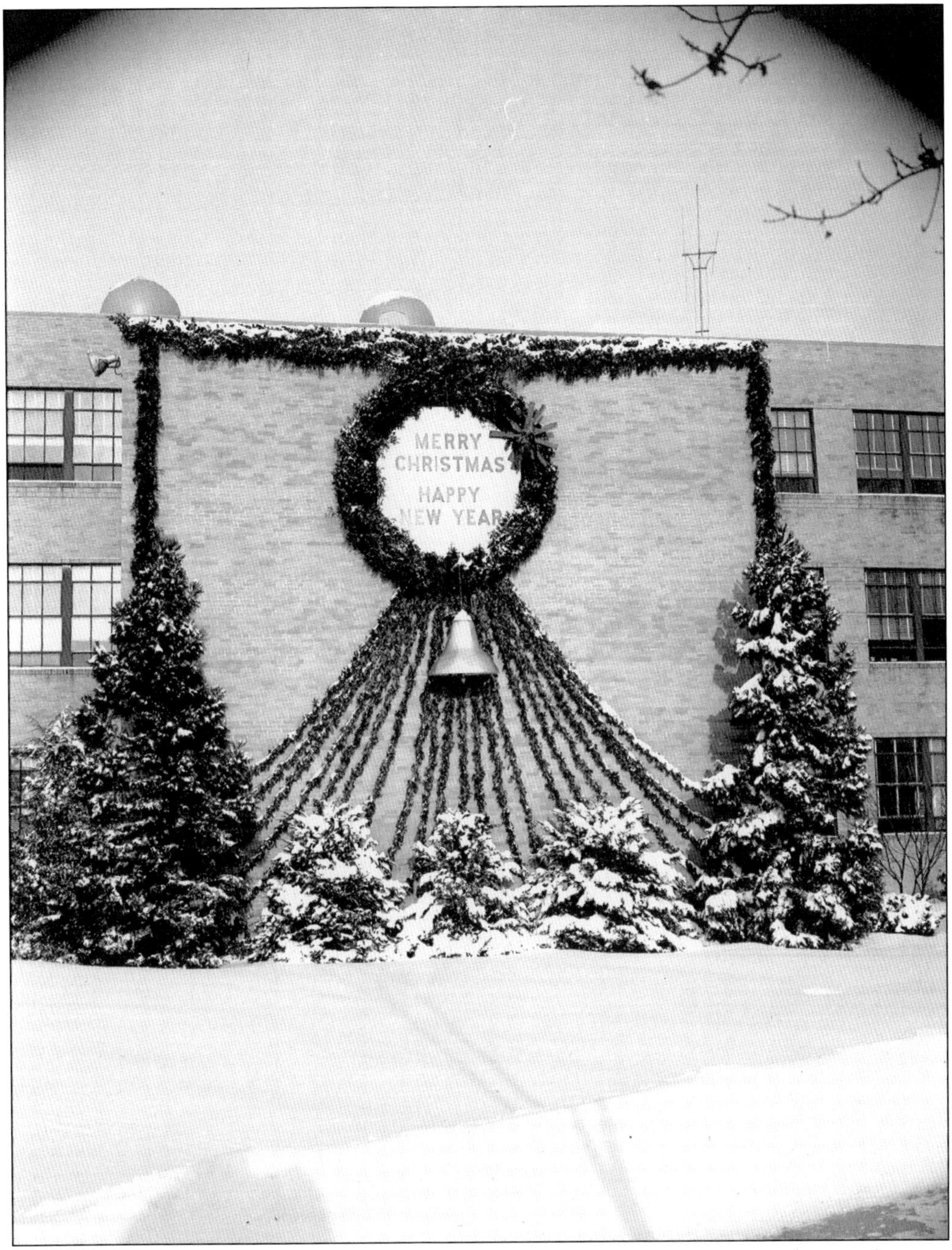

The facilities staff trimmed the side of the west wing for the holidays in December 1945.

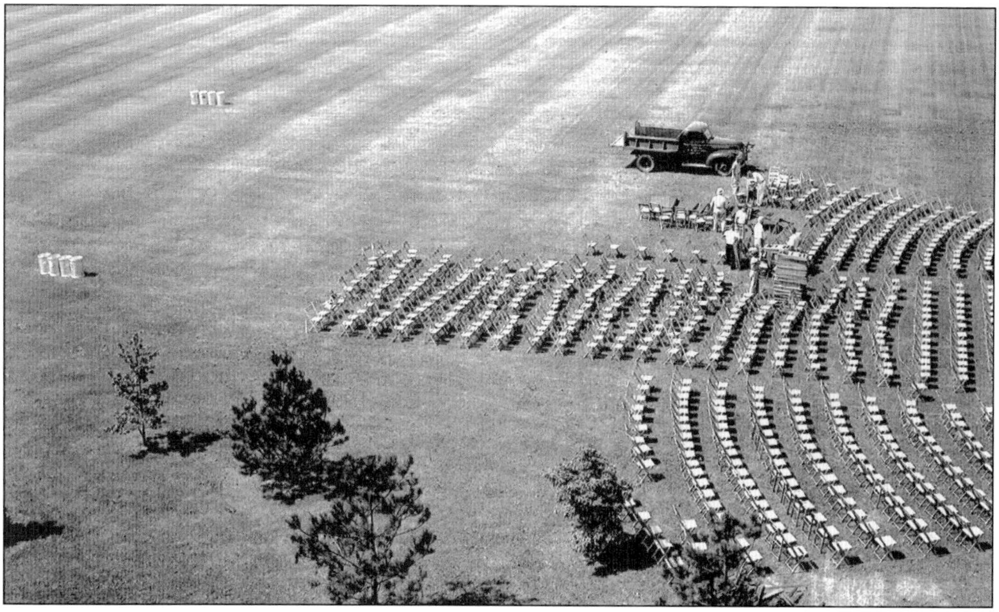

In June 1946, the facilities staff sets up for the labs' first annual picnic, which was to be held on the lawn in front of the west wing.

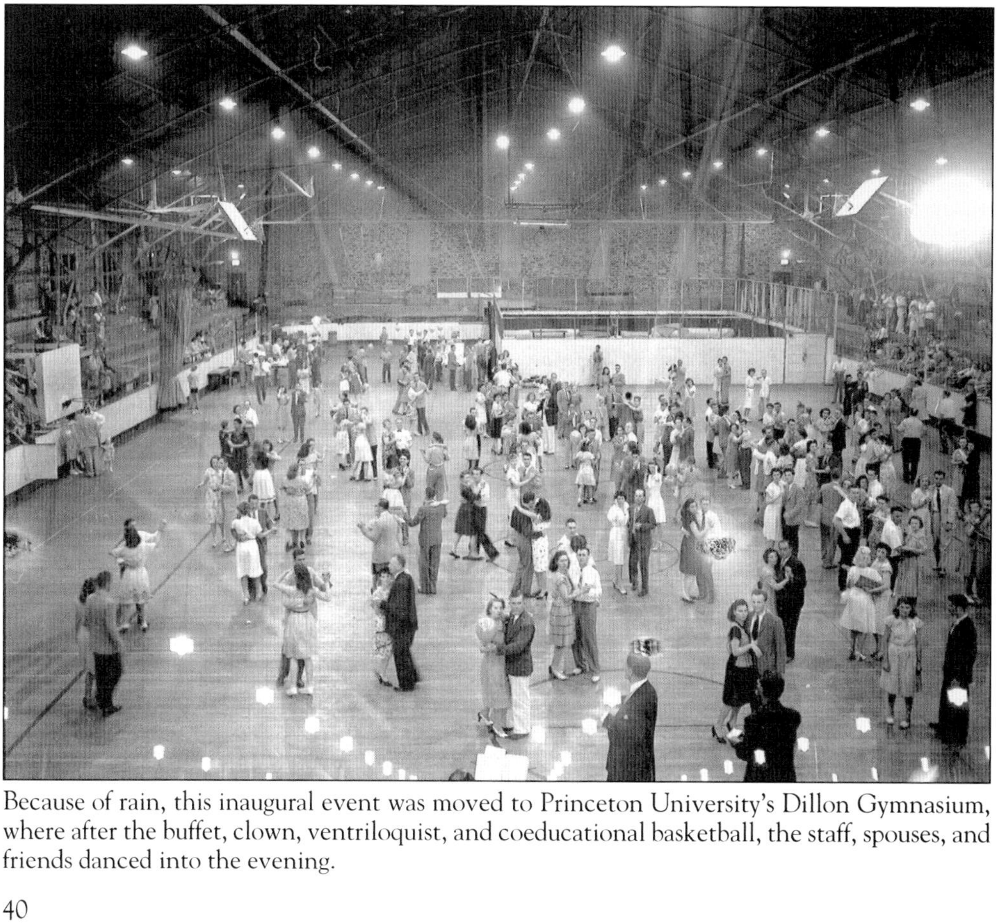

Because of rain, this inaugural event was moved to Princeton University's Dillon Gymnasium, where after the buffet, clown, ventriloquist, and coeducational basketball, the staff, spouses, and friends danced into the evening.

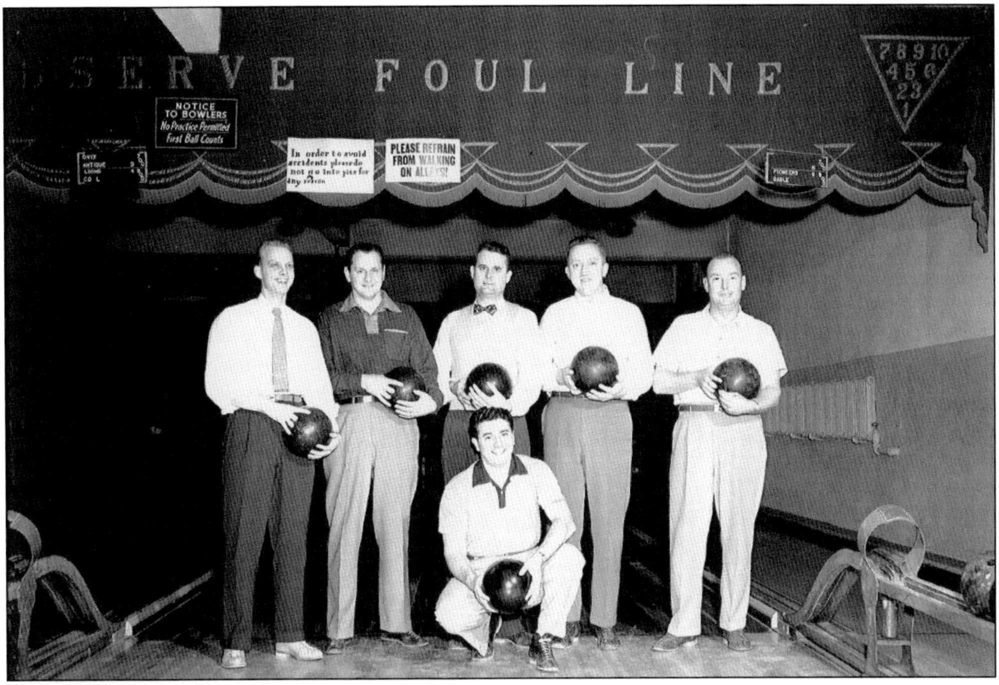

Pictured here are the finalists for the center's bowling championship in 1953.

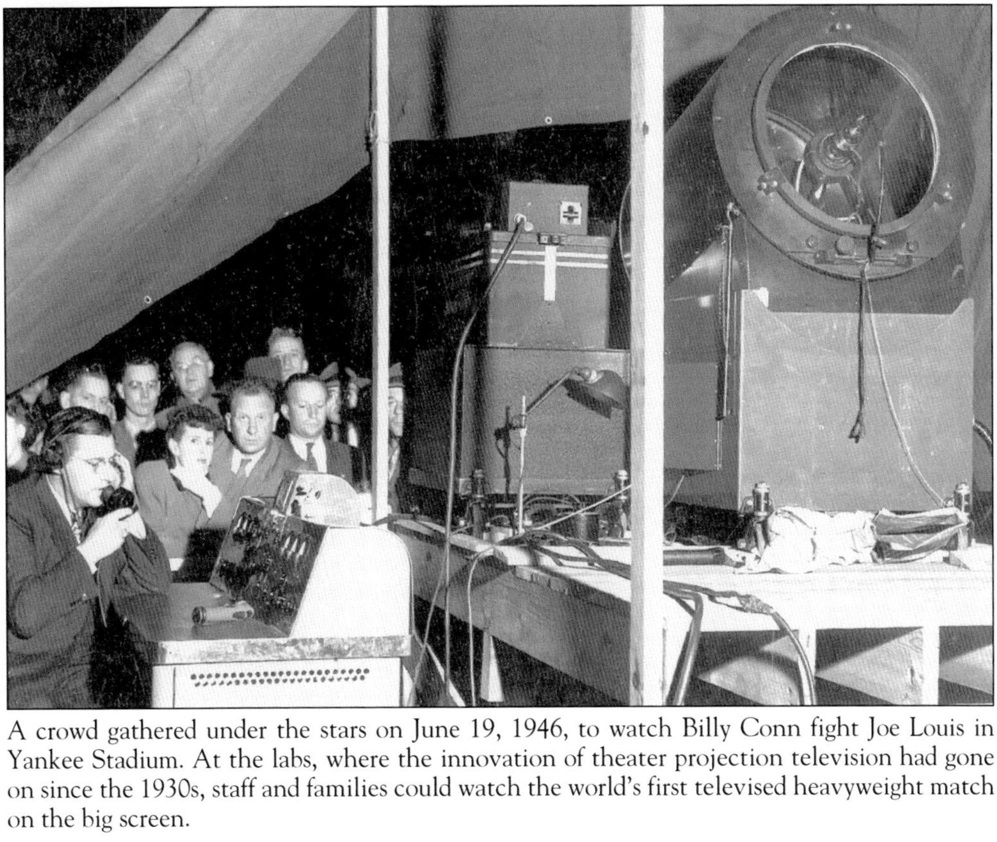
A crowd gathered under the stars on June 19, 1946, to watch Billy Conn fight Joe Louis in Yankee Stadium. At the labs, where the innovation of theater projection television had gone on since the 1930s, staff and families could watch the world's first televised heavyweight match on the big screen.

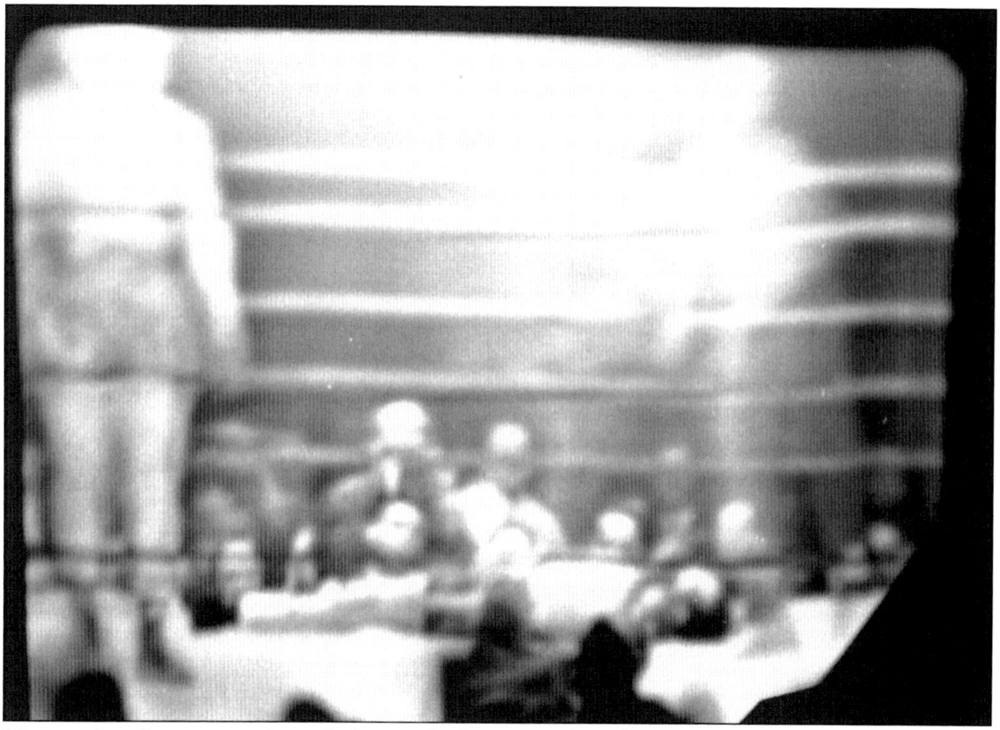

This is what they saw in the eighth round when Louis knocked out Conn.

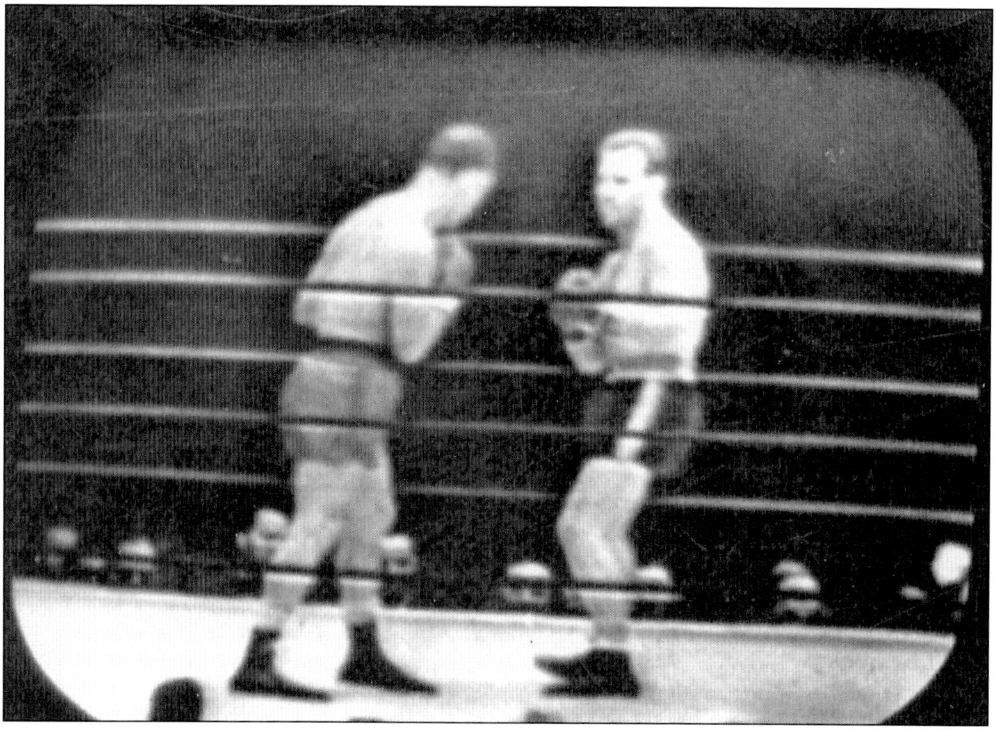

This is what RCA Labs engineer Les Flory saw on his standard 10-inch picture tube in Hopewell, New Jersey.

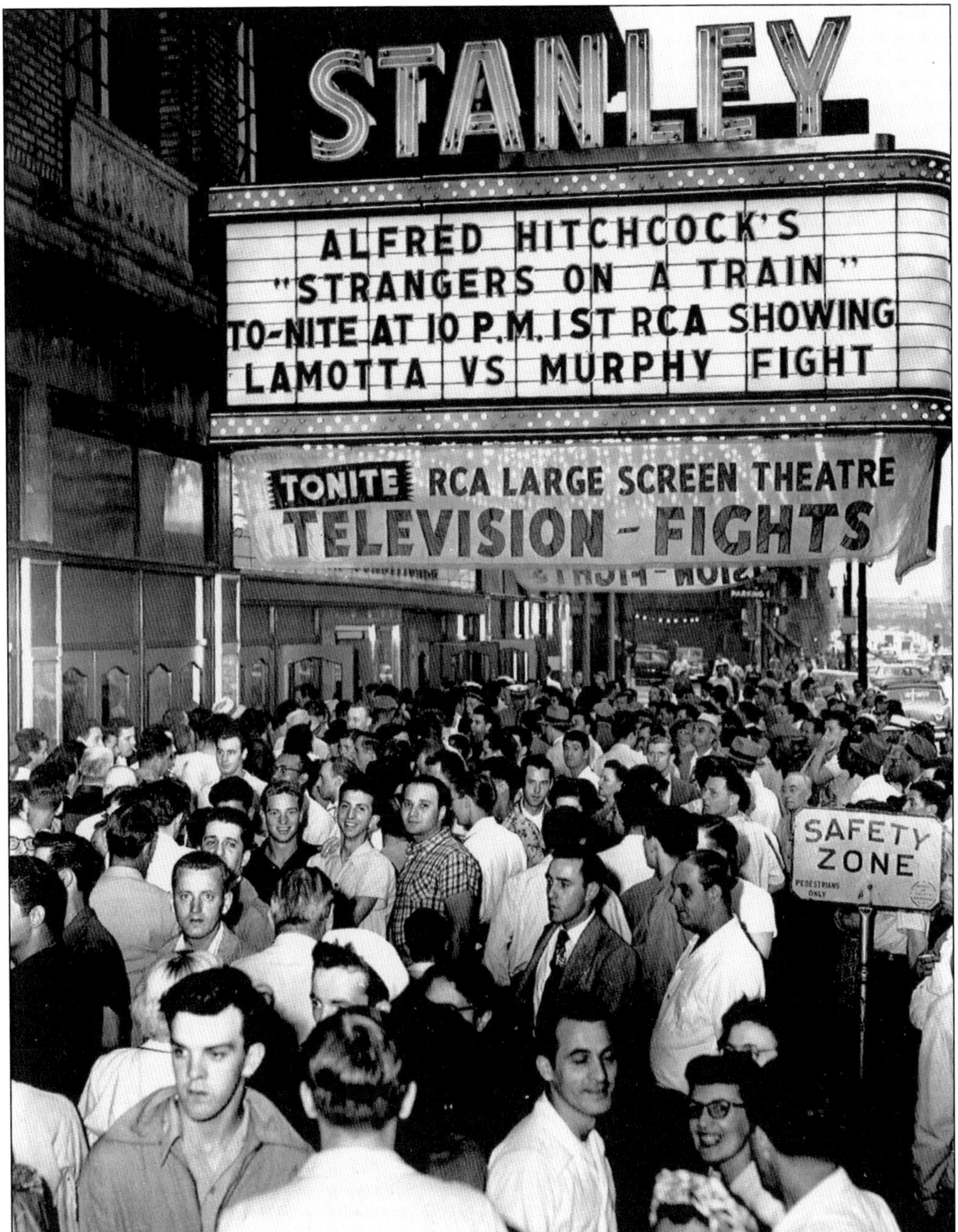

RCA promoted projection television to relieve the movie studios of fears that their audiences would stay home. This event was one of the last efforts at commercialization for 20 years, in Philadelphia in 1951.

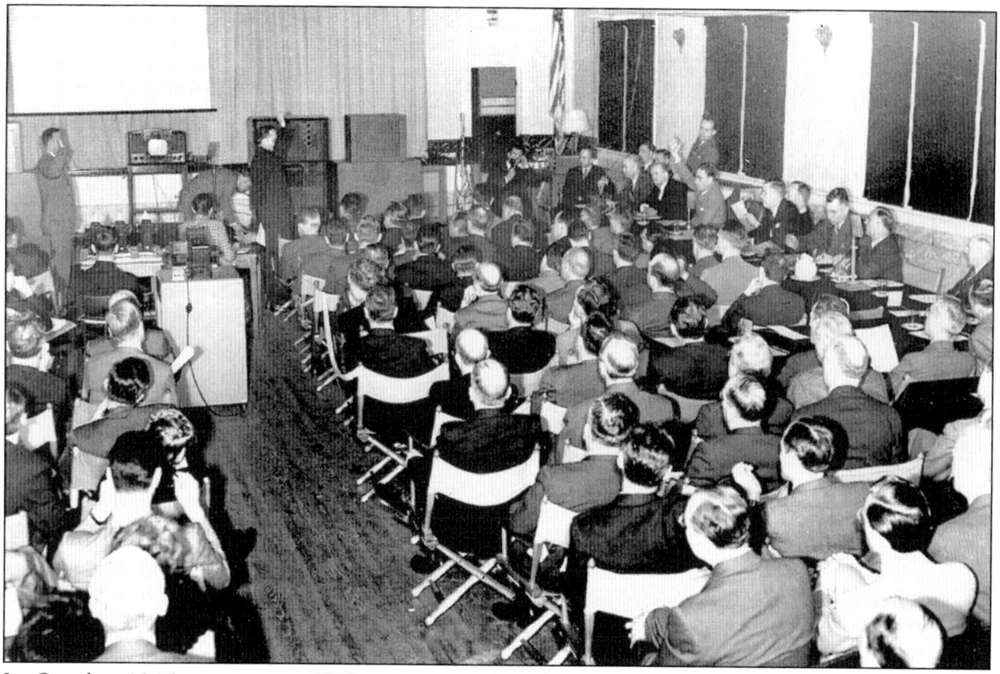

In October 1946, to counter CBS's argument that there was no alternative to color television besides its field-sequential system, RCA demonstrated all-electronic color television at the Penn's Neck Community Clubhouse, which has since been supplanted by the Alexander Road intersection with Route 1. Labs director Elmer Engstrom takes questions from the crowd of reporters while an array of RCA black-and-white and color sets lines the front of the room.

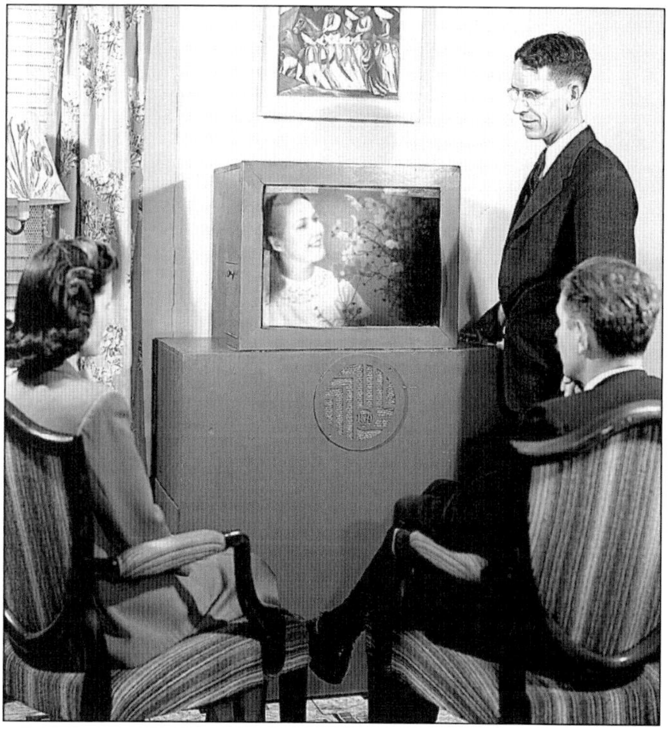

This couple might be watching color television, or they might be watching a transparency taped over the screen. At the cutting edge of invention, the timing of successful demos and the schedules of public relations do not always coincide.

George Brown, the RCA Labs' leader for electronic color television, tilts to the right outside the clubhouse. He is overseeing the erection of the antenna brought in to pick up the color television signal from NBC in New York.

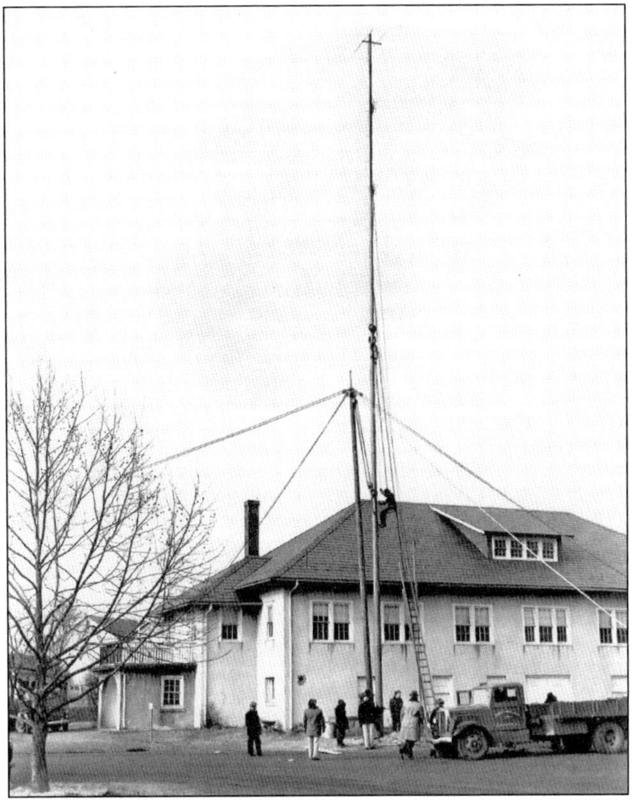

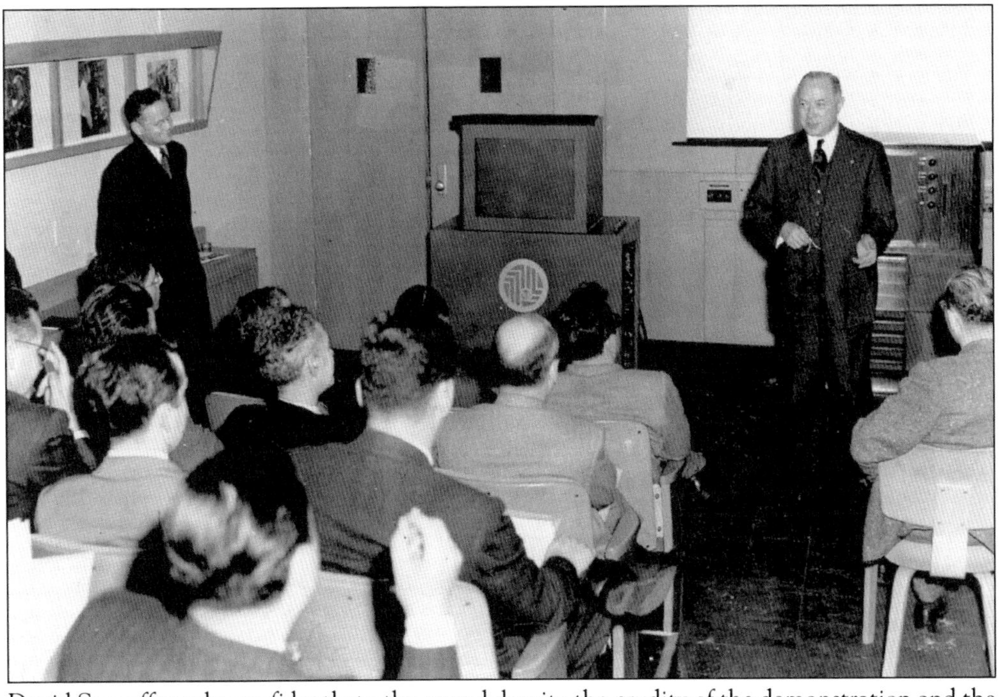

David Sarnoff speaks confidently to the crowd despite the crudity of the demonstration and the evident drawback of a television set the size of a small refrigerator.

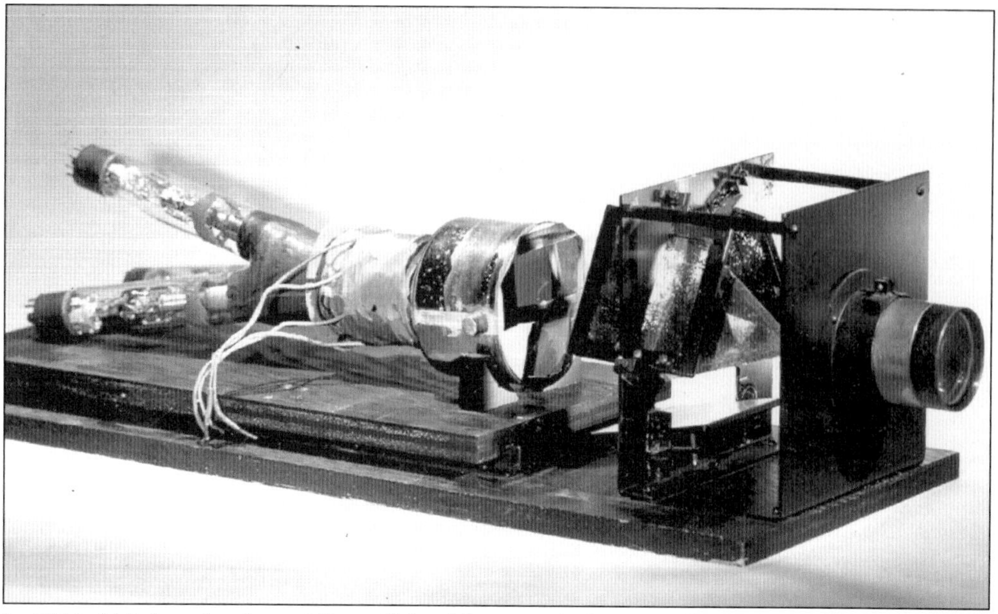

In 1947, Al Schroeder showed Sarnoff this "three-necked monster" at a private demo at the labs in Princeton. By feeding three electron beams through a common deflection yoke, he solved the problem of primary color registration on the screen. Sarnoff left fully convinced that electronic color television was a practical innovation. Schroeder also patented the shadow-mask picture tube that Harold Law reduced to practical fabrication in the fall of 1949.

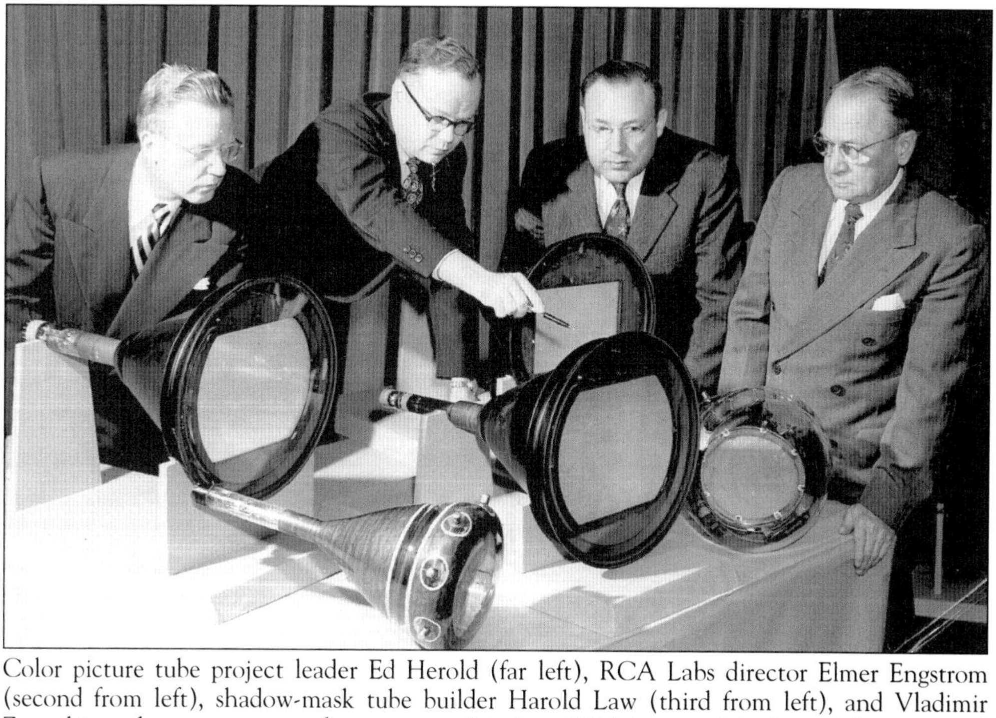

Color picture tube project leader Ed Herold (far left), RCA Labs director Elmer Engstrom (second from left), shadow-mask tube builder Harold Law (third from left), and Vladimir Zworykin gather at a press conference to explain how RCA invented, built, and demonstrated a color picture tube in six months in the winter of 1949–1950.

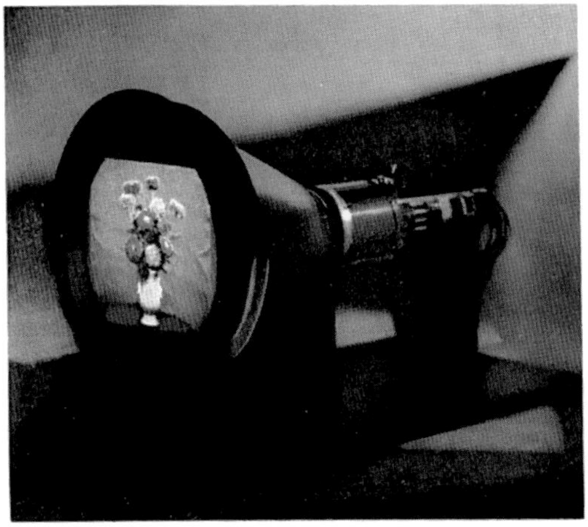

Research Report

1950

Just as in 1949, television was the first interest in 1950 for radio manufacturers and broadcasters. The extraordinary growth of television-receiver sales proceeded in spite of the new-station freeze, now more than two years old, and in spite of the confusion brought about by the color hearings during this beginning period for black-and-white television.

In research, we were concerned principally with color. As the year began we had finished our first presentation of the RCA color system to the Federal Communications Commission and we were preparing for a second comparative demonstration. We had just installed apparatus for automatically assuring color lock and had provided other improved circuitry. Demonstrations given included the comparative tests of February, the showing of the tri-color kinescope during late March and early April, demonstrations mainly to RCA personnel during August, and extensive demonstrations during December showing current status.

When the hearing closed, near the end of May, we believed we had made such a showing as to give assurance of operation, satisfactory performance, and compatibility. Throughout the year we made progress on performance refinements and apparatus simplifications.

i

Never before and never after did the RCA Laboratories corporate annual report feature color. This was the exception, as was the image of an apparently working color tube. RCA's legal department forbade color photographs of the new sets on the grounds that the reality might not look as good as what appeared in the press.

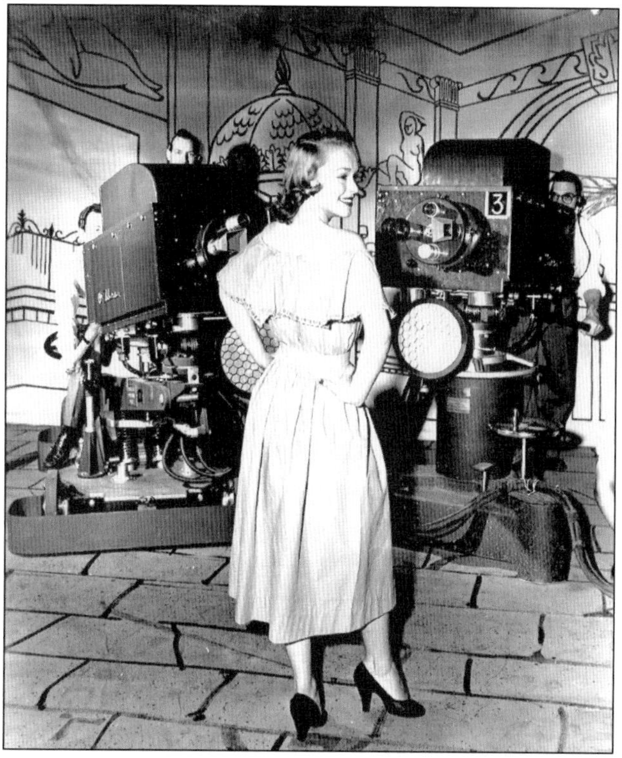

Ann Schaeffer strikes a pose for the color television cameras at the Colonial Theatre in New York. There, RCA's subsidiary, the National Broadcasting Company (NBC), staged productions in 1951 for field tests of the color television system based almost entirely on RCA Labs' proposals.

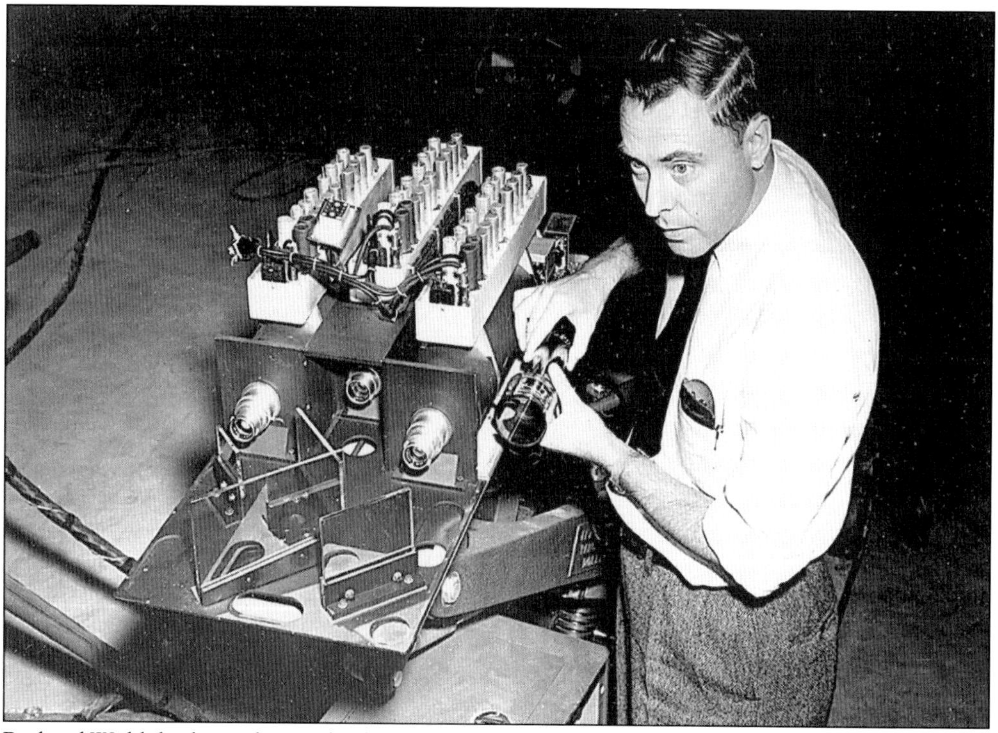

Richard Webb looks up during the field tests from the image orthicon tubes and dichroic mirrors that make up the heart of the color television camera he invented.

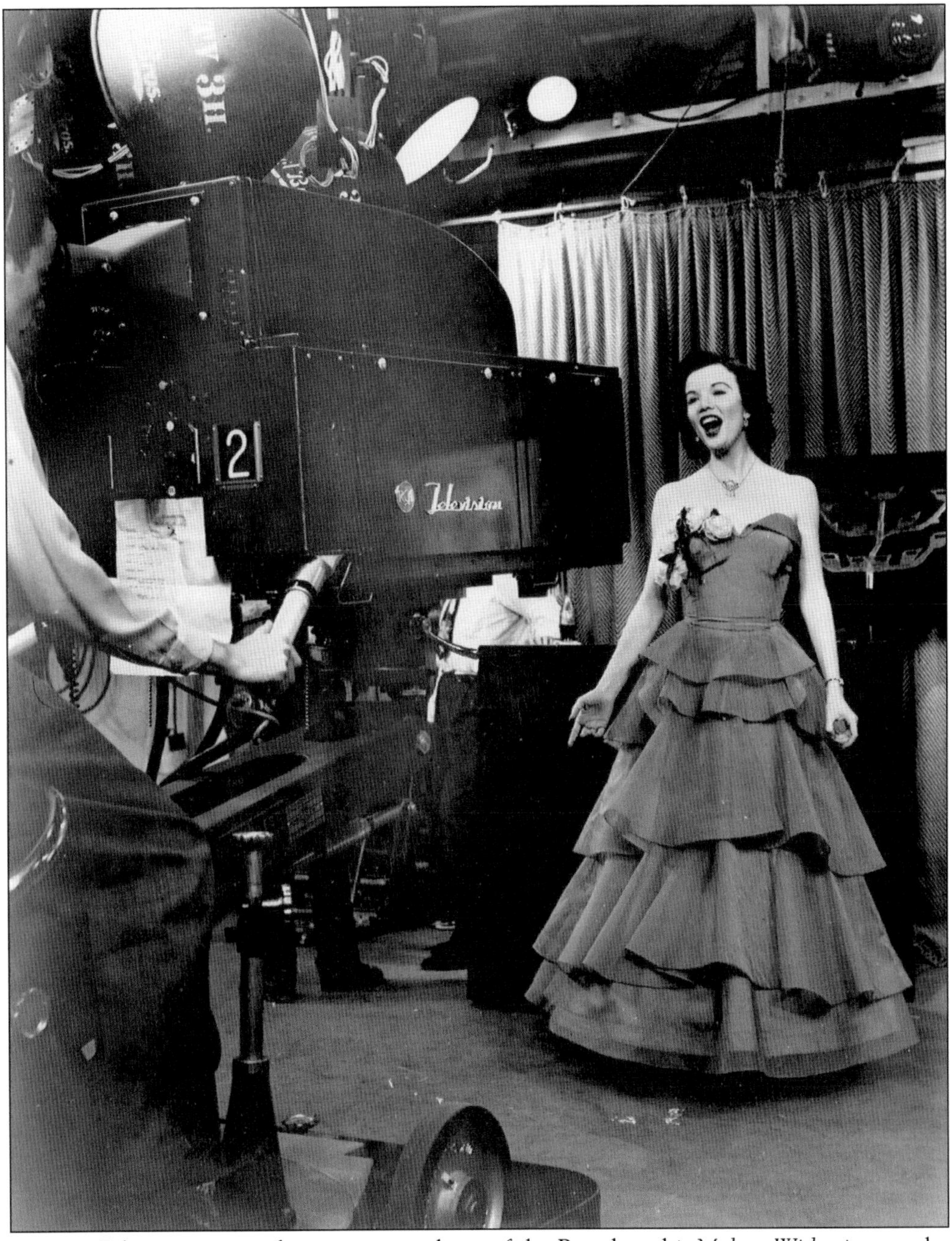

Nanette Fabray, mistress of ceremonies and star of the Broadway hit *Make a Wish*, sings to the color cameras and audience in June 1951. Virtually everyone watched her in black-and-white because the RCA system was compatible with the earlier broadcast standard.

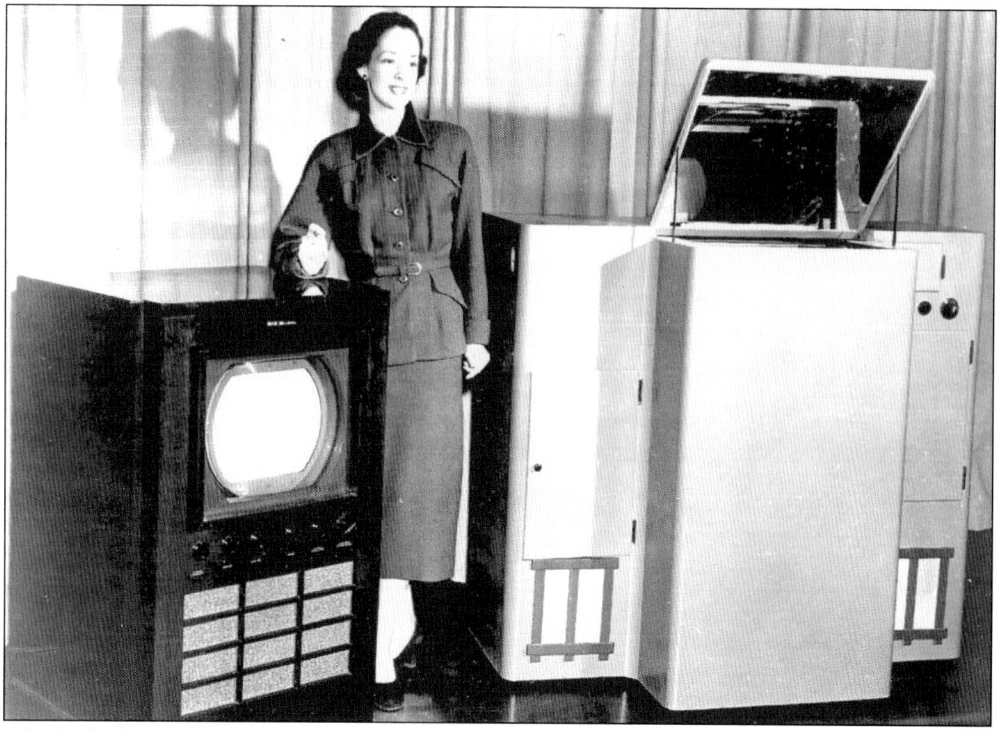

RCA's shadow-mask picture tube television is being compared to the "Trinoscope" in December 1950.

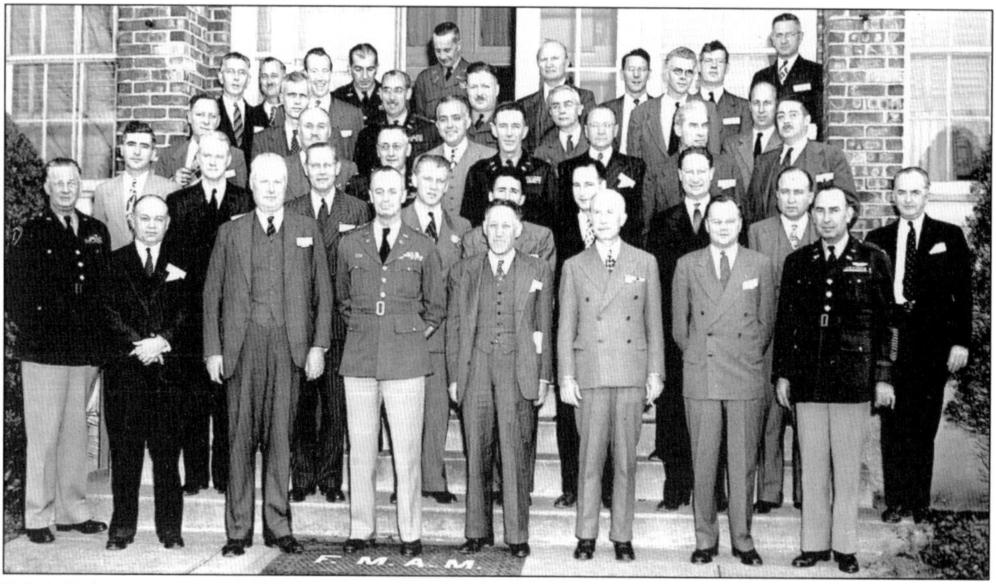

RCA Laboratories and RCA Victor staff members meet with the Signal Corps staff at Fort Monmouth, New Jersey, for two days' discussion of research-and-development projects late in 1948. Harold Beverage, Chief Signal Officer Akin, Irving Wolff, M.C. Batsel, RCA Labs director Elmer Engstrom are third through seventh from the left in the front row. Harry Olson is second from the right in the second row, and Vladimir Zworykin stands third from the right in the third row.

The world's first digital electronic memory stored and rewrote 256 bits of information based on the control of electron beams and the phenomenon of capacitance. Designed in 1946 by the labs' Jan Rajchman for John von Neumann's programmable, electronic, general-purpose computer at the Institute of Advanced Study, the Selectron was later used in Rand Corporation's "Johnniac" through the 1960s. Handmade at RCA's factory in Lancaster, Pennsylvania, Selectrons sold for $8,000 apiece.

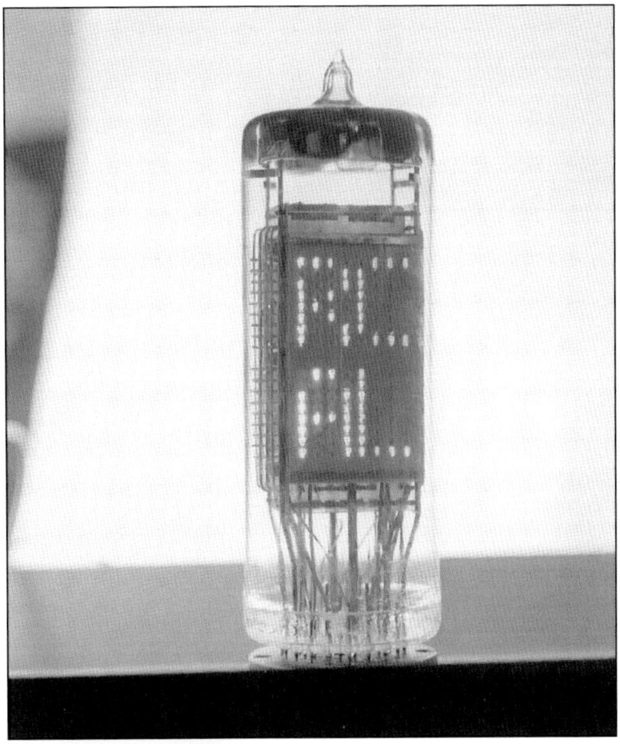

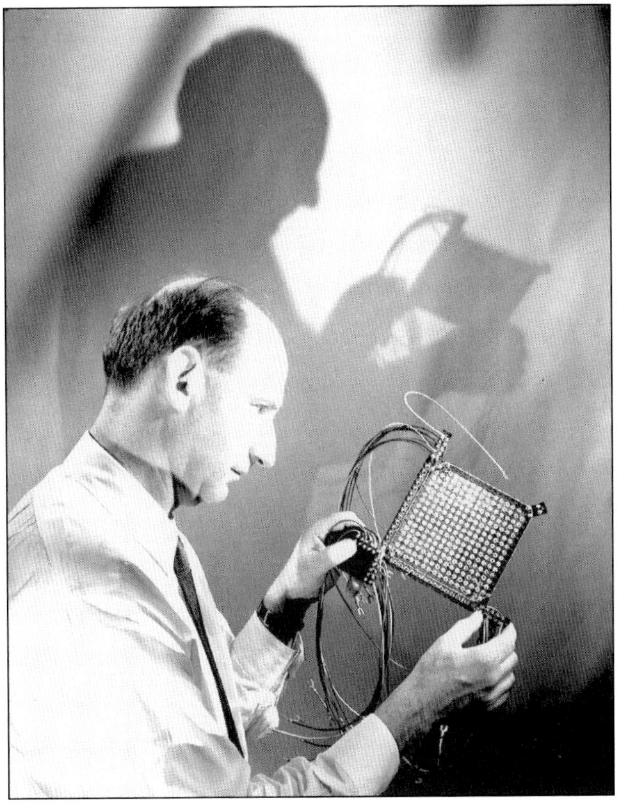

Rajchman holds one of the first magnetic core memories containing an array of 18-by-18 hand-strung spools that read north or south depending on the flow of current. Computers used memories based on this technology for the next 25 years.

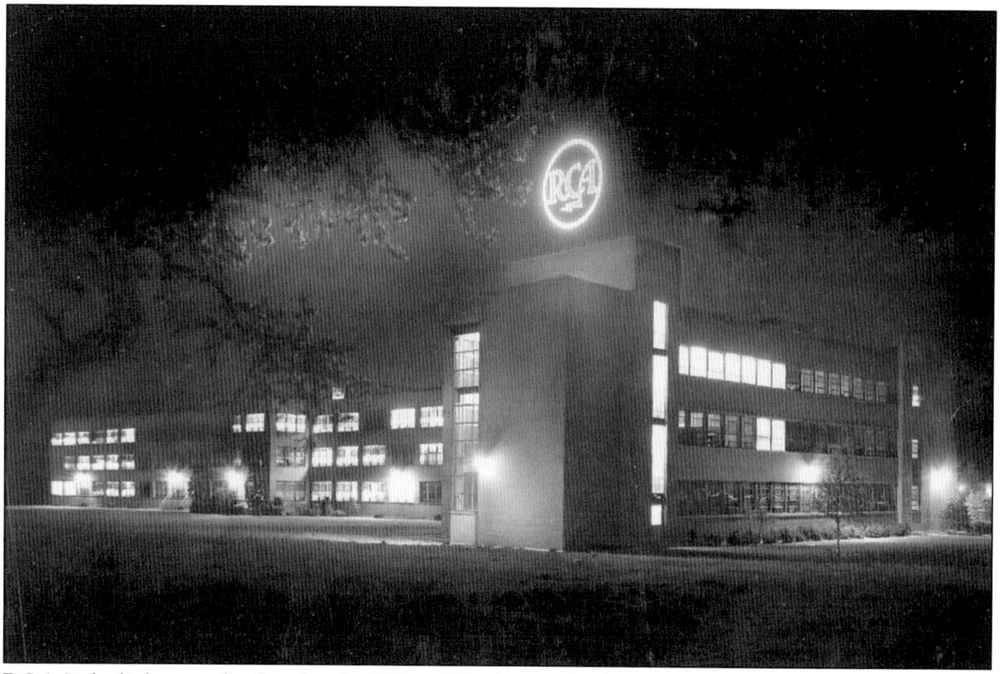

RCA Labs lights up the farmland of West Windsor with the "meatball" logo atop the new west wing extension in 1948.

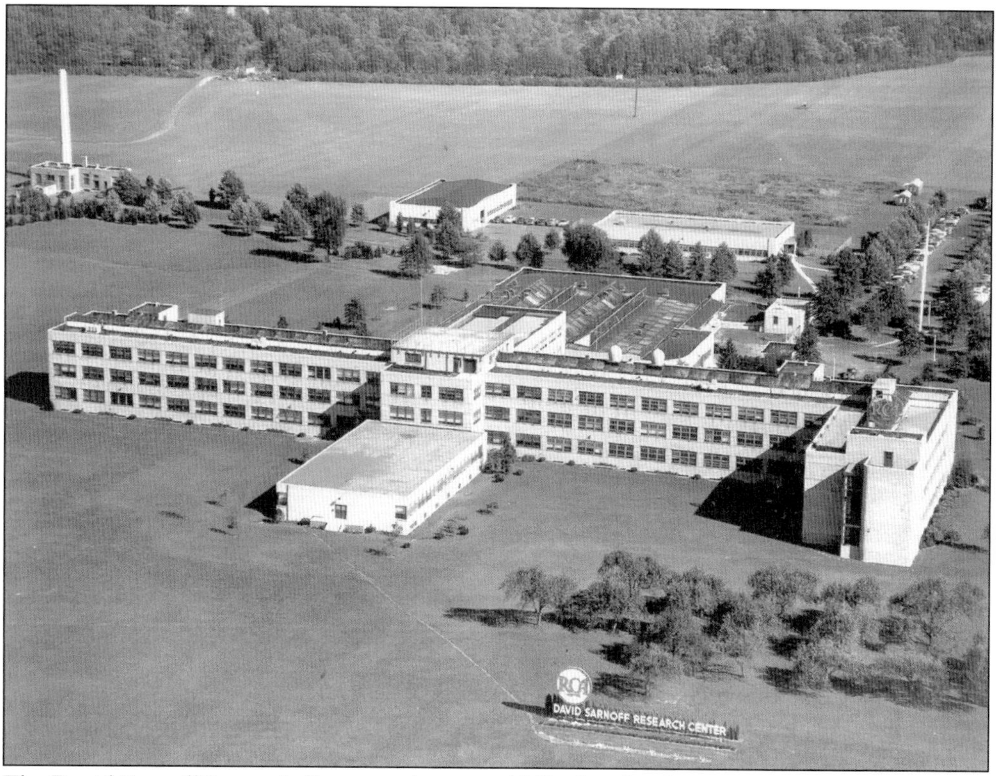

The David Sarnoff Research Center is shown in 1952, shortly after it was renamed.

On September 27, 1951, RCA chairman and chief executive officer David Sarnoff was feted on his 60th birthday at RCA Laboratories. After lunch in the cafeteria, a bronze plaque was unveiled, honoring his 45 years in electronics and renaming the Princeton facility the David Sarnoff Research Center. The plaque now hangs in the hallway next to Sarnoff Corporation's lobby.

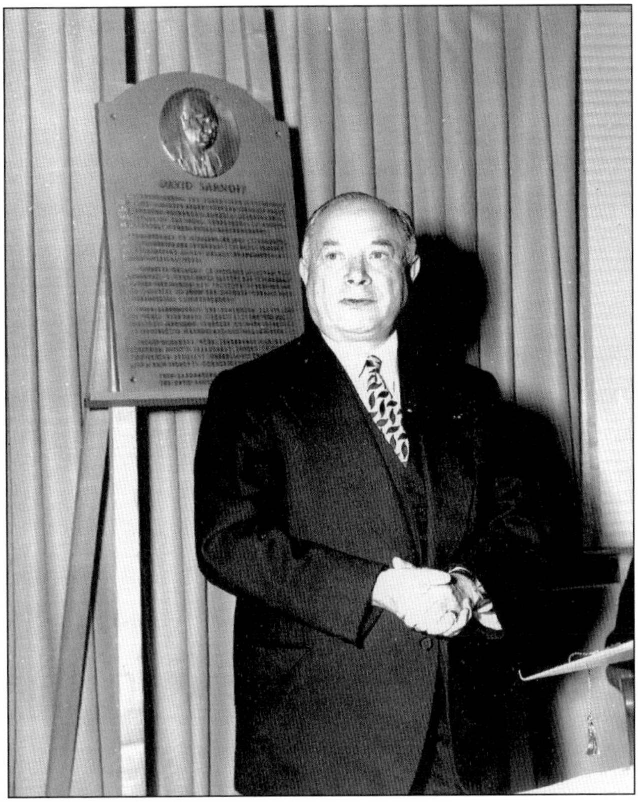

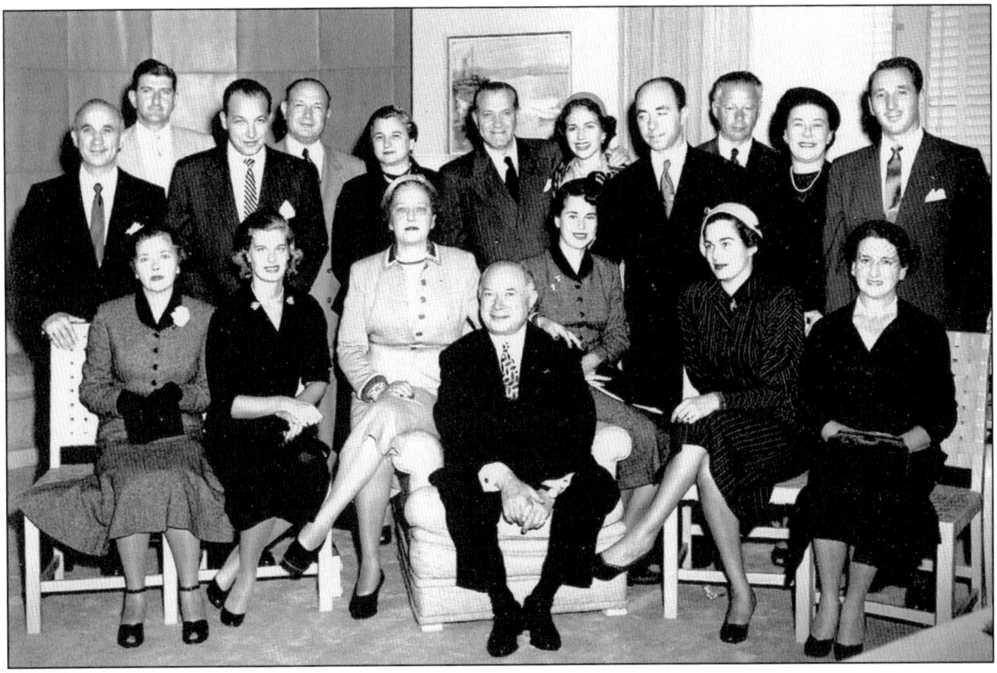

Much of David Sarnoff's family gathered for the occasion.

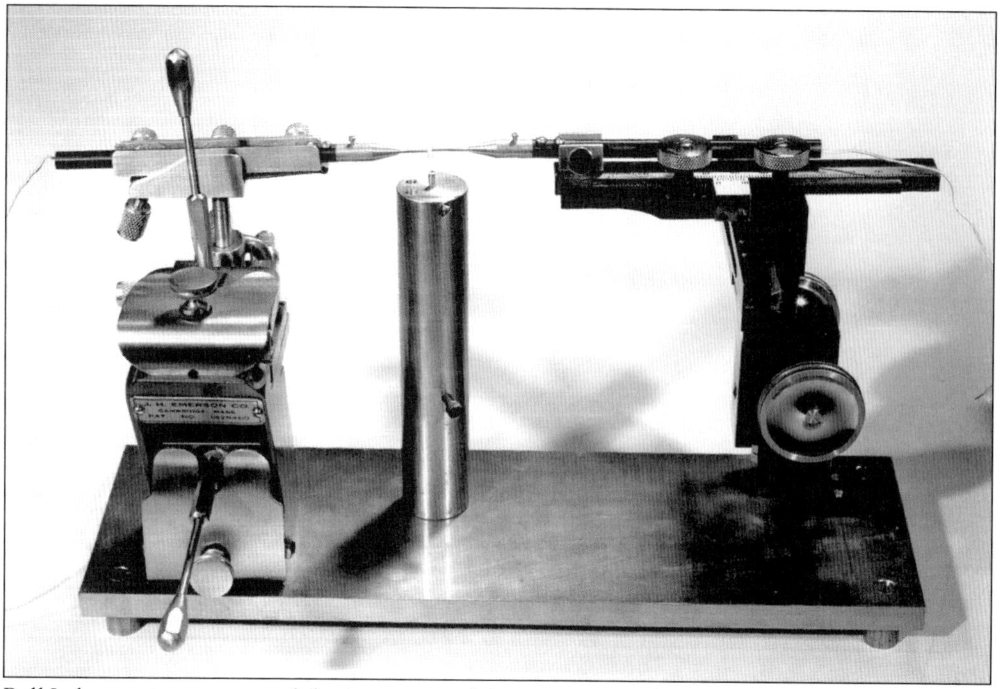

Bell Laboratories announced the invention of the transistor in Murray Hill, New Jersey, in June 1948. Less than two weeks later, Jerome Kurshan at the labs assembled RCA's first device based on quantum electronics.

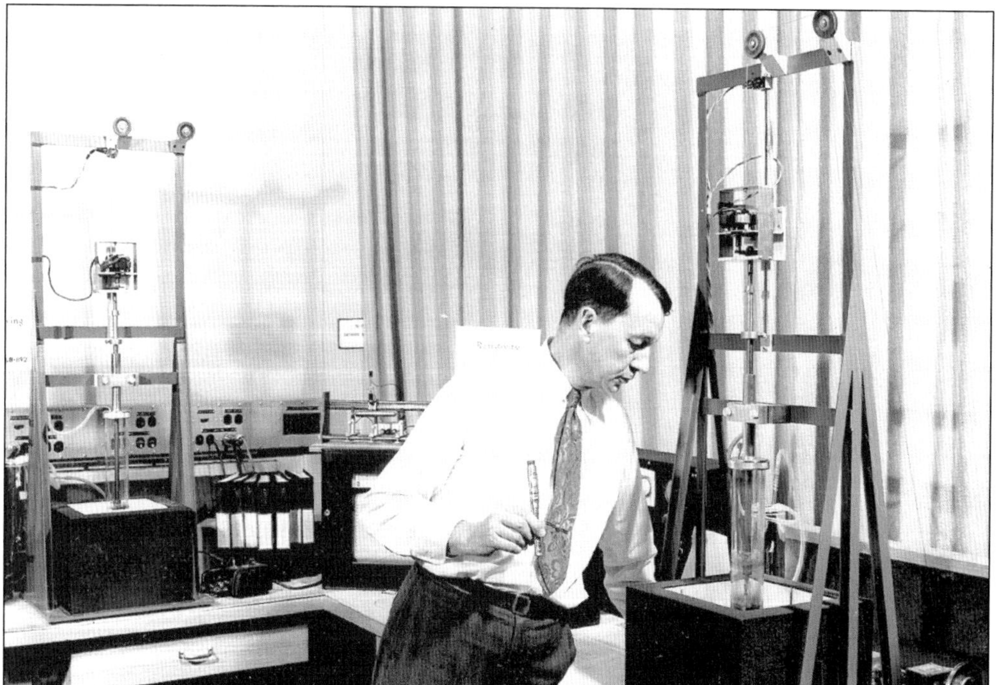

Schuyler Christian shows off the germanium crystal pulling facilities at the labs, still the basis for creating the pure materials necessary for semiconductor devices.

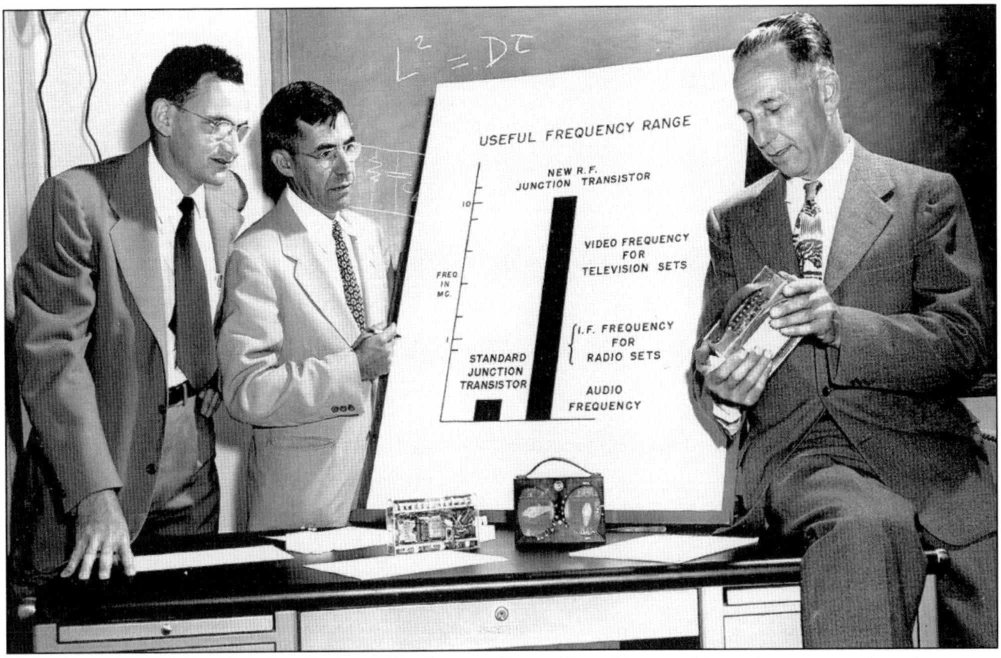

Jacques Pankove (left), Charles Mueller (center), and Loy Barton show the tremendous increase in frequency range attained with their junction transistors in 1952. Previously practical only for hearing aids, the transistor could now be applied to a wide range of portable devices.

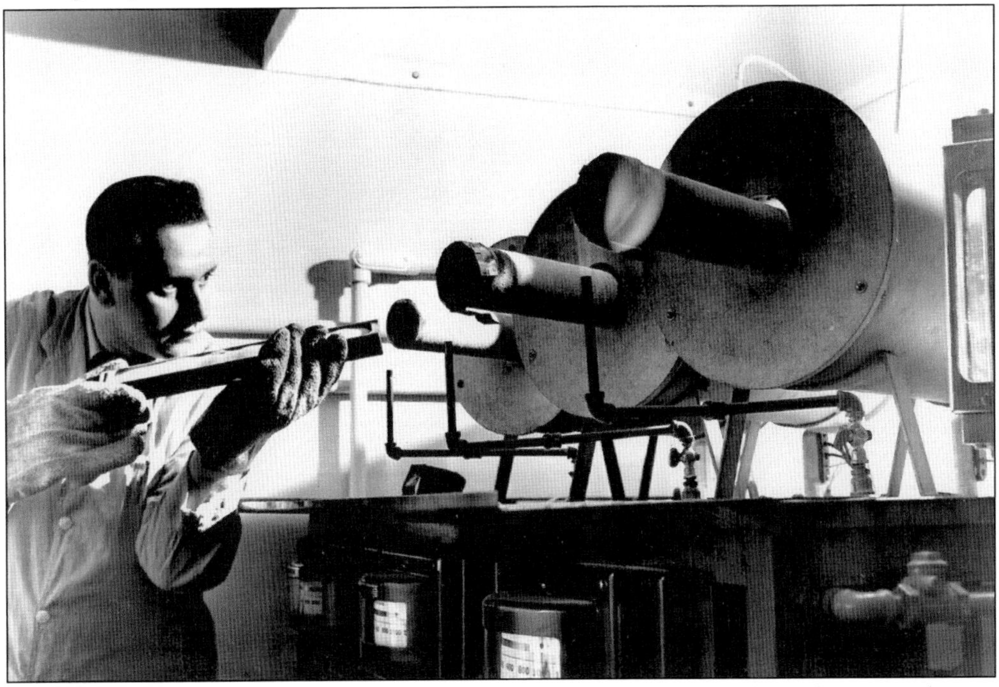

A technician loads trays of germanium powder into hydrogen furnaces heated to 1,000 degrees Centigrade at RCA Victor's electron device factory in Harrison, New Jersey. Staff members at Harrison and Princeton were exploring methods of making pure germanium without pulling crystals.

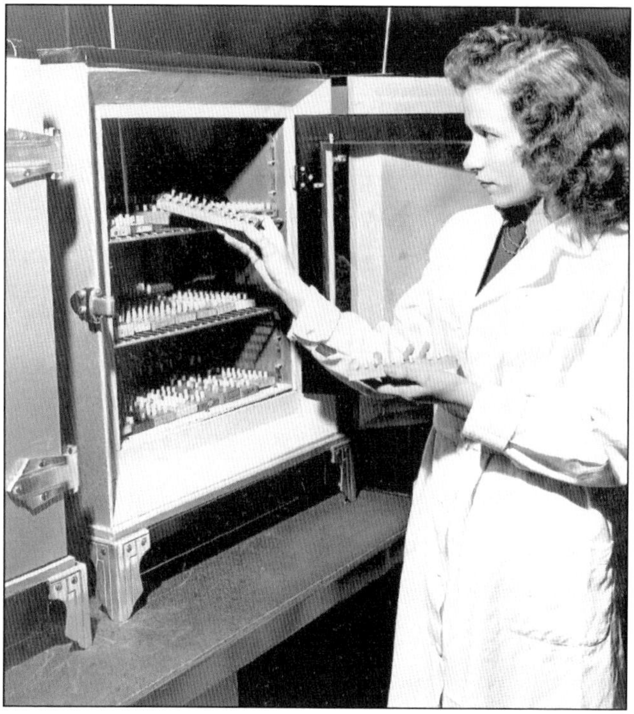

Another Harrison technician loads trays of plastic-coated transistors in an oven for three days of baking before they are tested for power output.

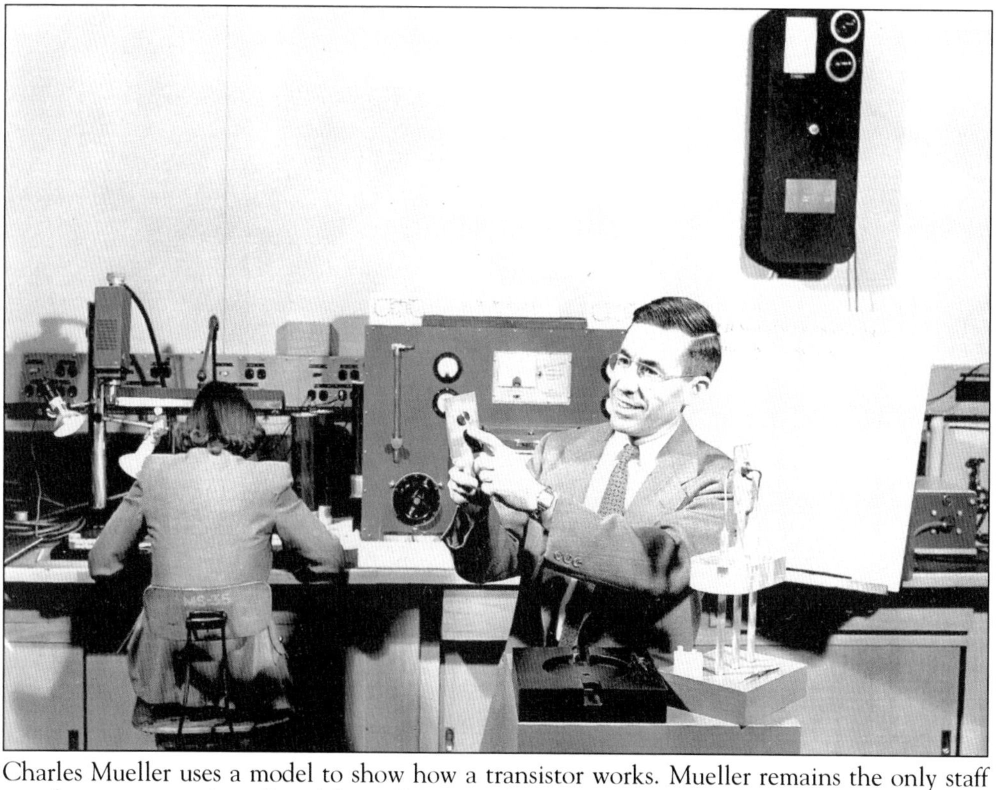

Charles Mueller uses a model to show how a transistor works. Mueller remains the only staff member to receive three David Sarnoff Outstanding Achievement Awards.

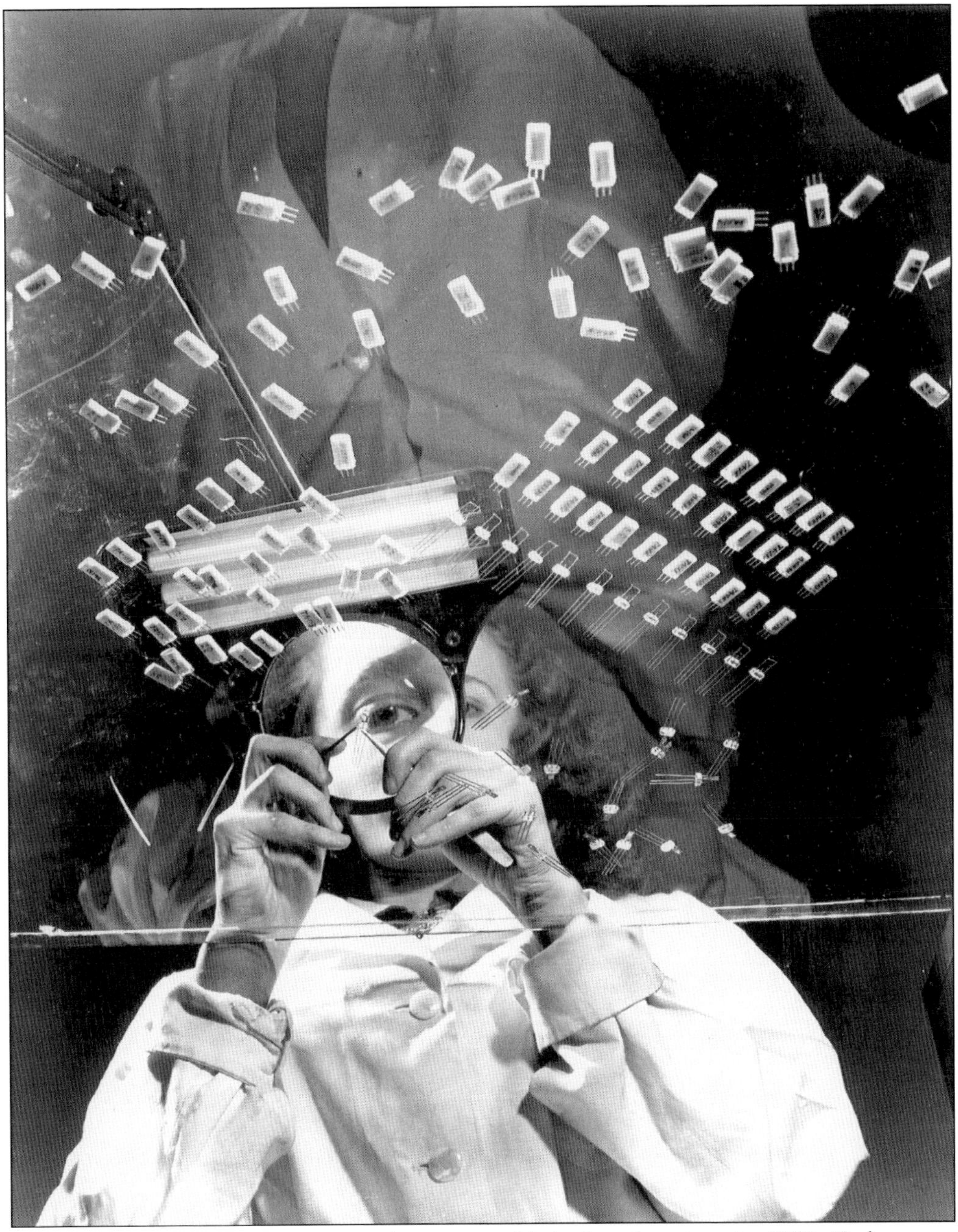

Here we see transistor components being assembled piece by piece in 1953. Today, over 50 million transistors can be put on a computer chip.

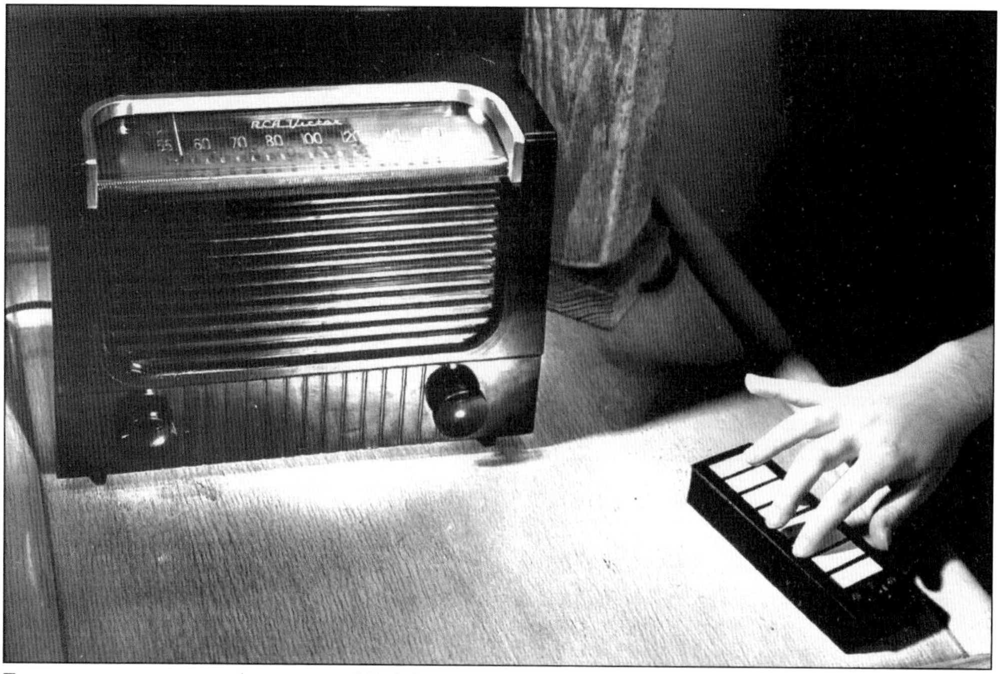

For a presentation to licensees of RCA patents in November 1952, the staff built a variety of devices to show how transistors could reduce their weight and power requirements. This keyboard can be heard through the radio's speaker by wireless transmission.

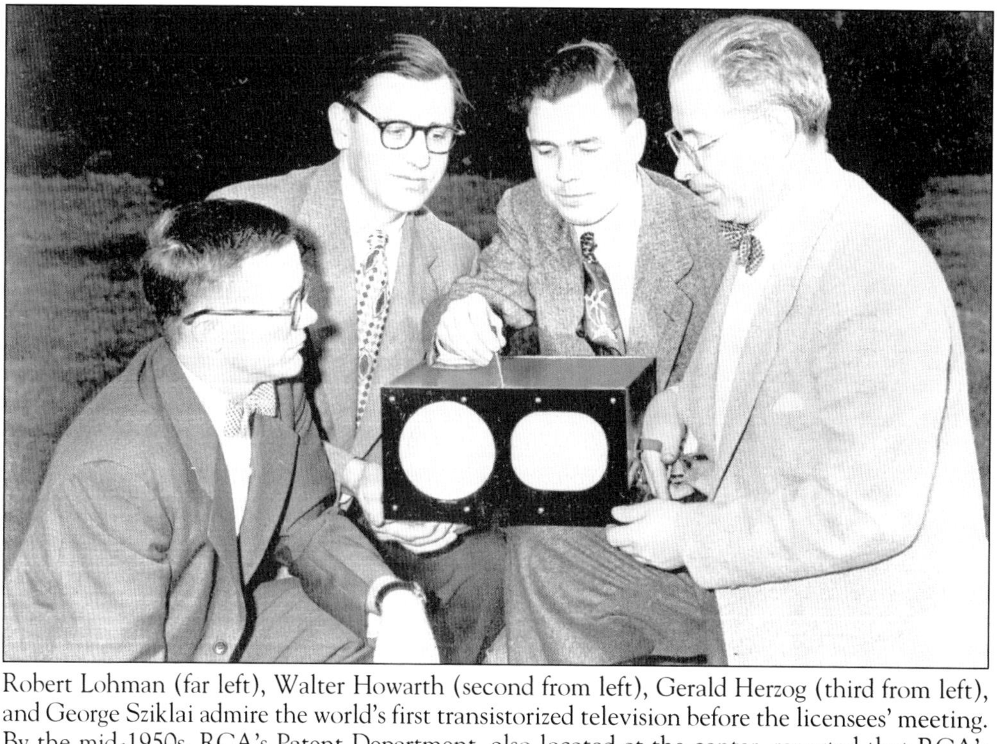

Robert Lohman (far left), Walter Howarth (second from left), Gerald Herzog (third from left), and George Sziklai admire the world's first transistorized television before the licensees' meeting. By the mid-1950s, RCA's Patent Department, also located at the center, reported that RCA's patent position in transistors was second only to that of the much larger Bell Laboratories.

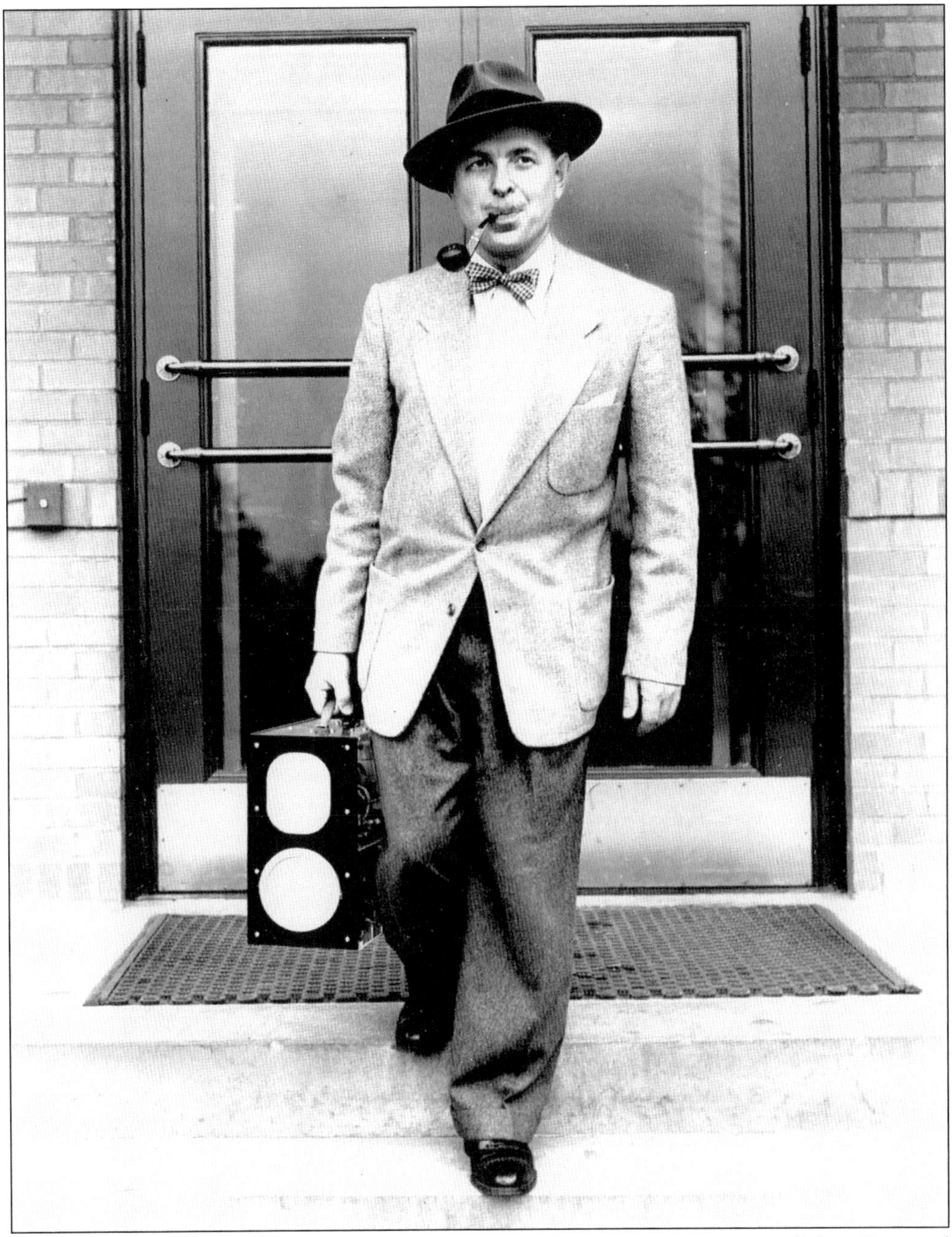

George Sziklai steps purposefully down the steps of the center's entrance with his 27-pound portable transistorized television. Good reception was obtained up to five miles from the Empire State Building. With further tweaking, the group pushed it to 15.

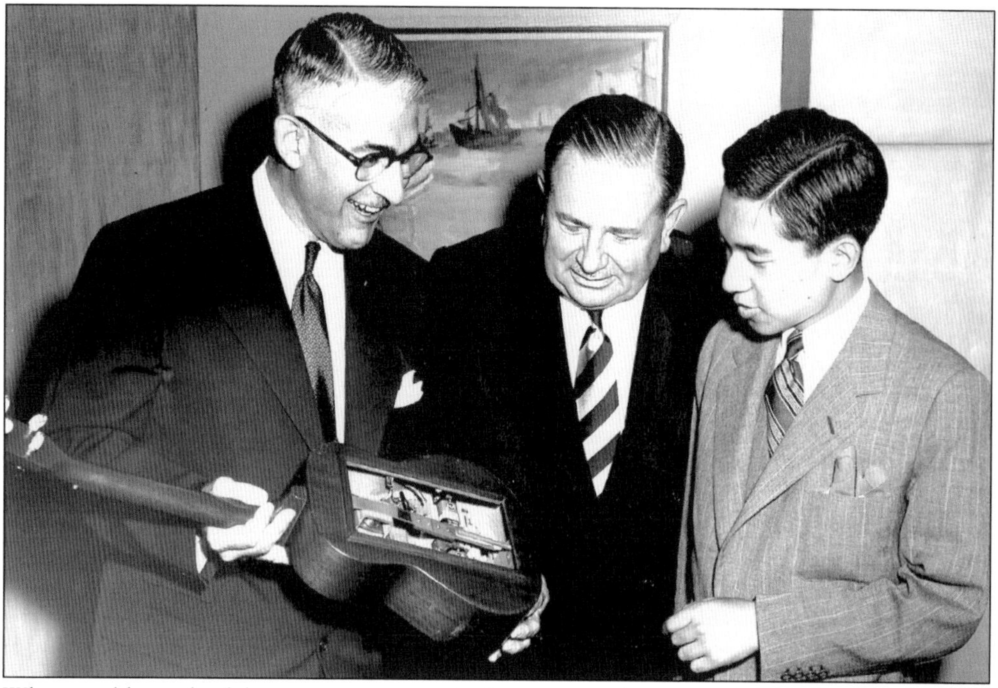

What would you do if the crown prince visited your research center? Cyril Hoyler shows RCA president Frank Folsom and Japan's Prince Akihito the delights of a solid-state ukulele in 1953. The instrument was a fad at the time because Arthur Godfrey played one on his NBC variety hour.

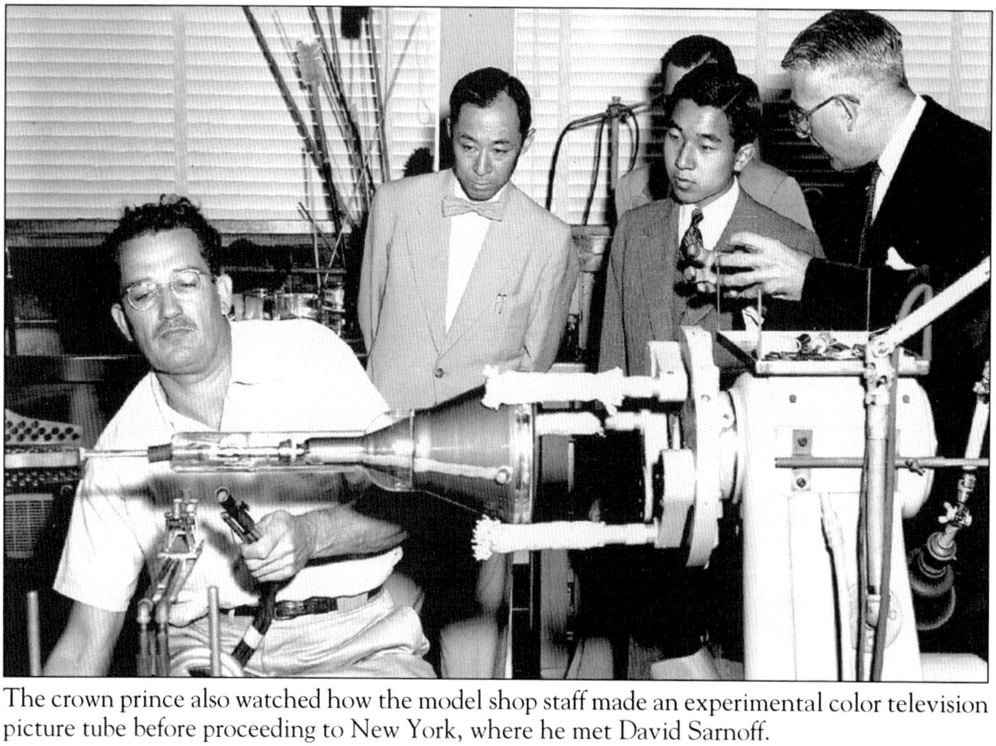

The crown prince also watched how the model shop staff made an experimental color television picture tube before proceeding to New York, where he met David Sarnoff.

Three

Country Club or Golden Age? 1954–1971

As RCA expanded in the prosperous postwar era to more than 30,000 employees in New Jersey, and 100,000 worldwide, so too did the David Sarnoff Research Center. This was an era of spectacular industrial research into the properties of elements and alloys, the conversion of energy from one form to another, the reduction of transistors in size and power demands. It gave rise to the center's first Nobel laureate, solar cells, thin-film and MOS transistors, CMOS integrated circuits, and liquid-crystal displays.

Much of this research drew on funding from military agencies looking a generation ahead for new technologies. It also benefited from income on patents licensed to Japanese and European companies after an antitrust settlement in 1958 gave RCA's 3,000 color television patents to domestic competitors. As a result, many people in RCA's product divisions called the landscaped Princeton campus a "country club" staffed by scientists with no experience in the realities of factory engineering. Yet those researchers' explorations and innovations, sometimes barely ready for their breadboard demos, enabled the handheld and wireless technologies of today.

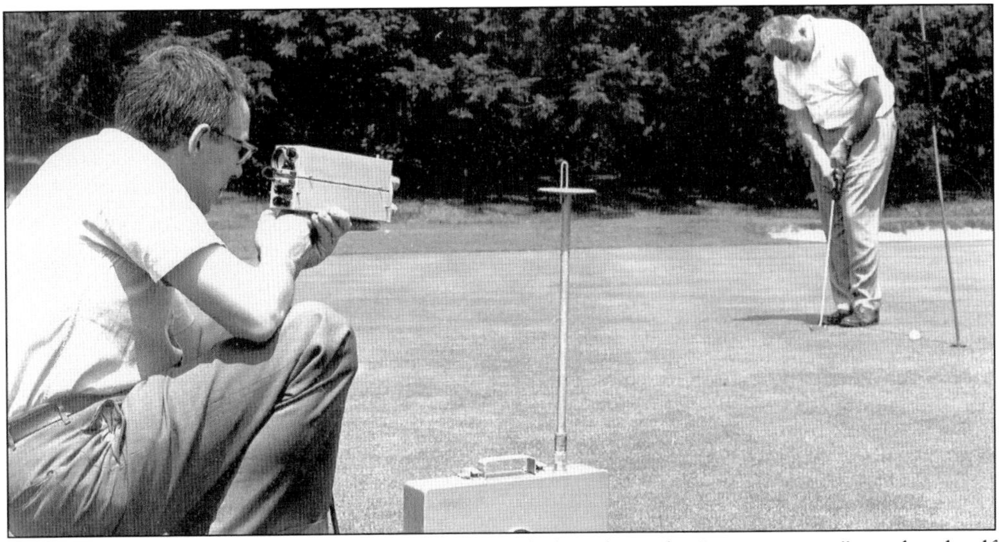

Win Pike answered the question in the chapter title by taking the "creepy peepy" to a local golf course to demonstrate its transmission range and portability. This and other nonbroadcast video cameras were all based on the vidicon tube invented by Paul Weimer in the late 1940s.

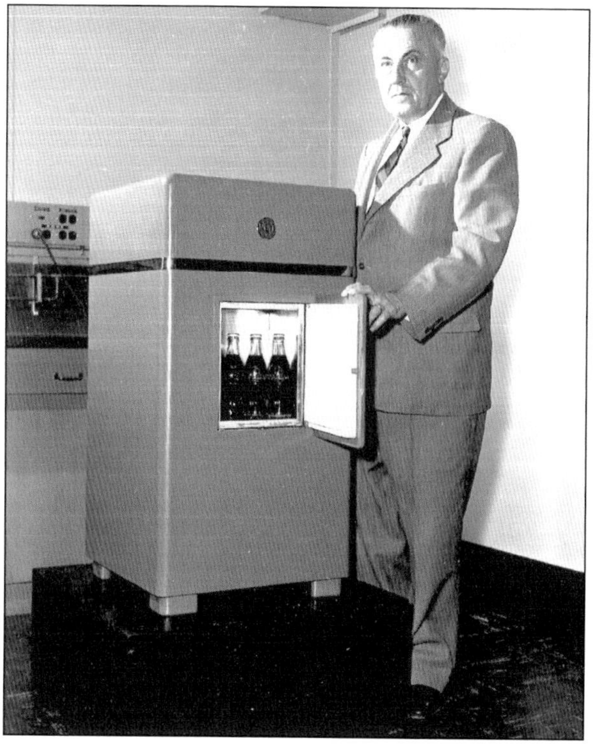

For his 50th anniversary in electronics (in 1956), David Sarnoff requested three presents from the center's staff. One was noiseless electronic cooling because RCA had recently purchased the Whirlpool Corporation. Nils Lindenblad responded by using the Peltier effect to eliminate the need for motors or compressors. Instead, one needs lots of surface area for heat exchange, which explains why this refrigerator is so small.

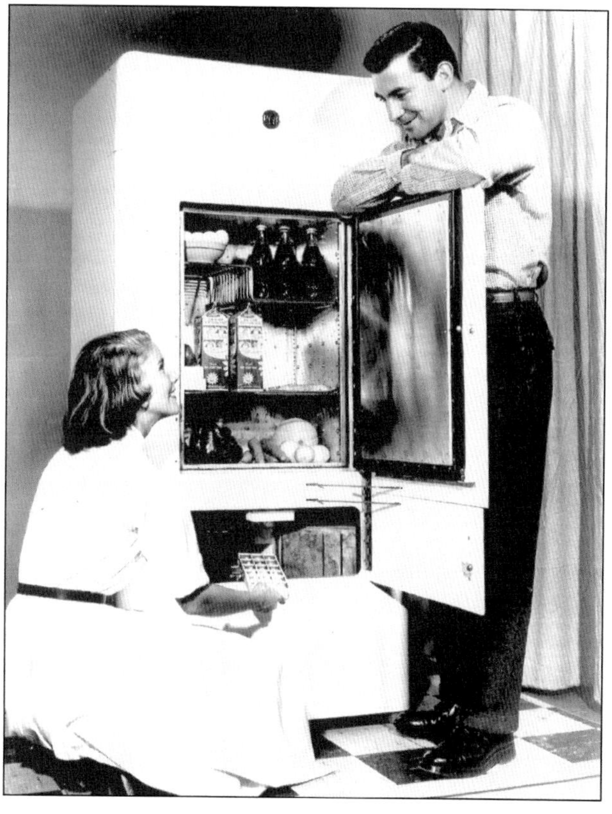

When he asks, "Honey, what's in the fridge?" she could respond, "The most expensive ice cubes in the world." The labs' George Brown volunteered that comment at a professional meeting where General Electric researchers reported on the first man-made diamond. Note the local dairy purveyor of milk, headquartered in nearby Plainsboro.

She admires the fins on the back of the panels used to cool or heat the room inside.

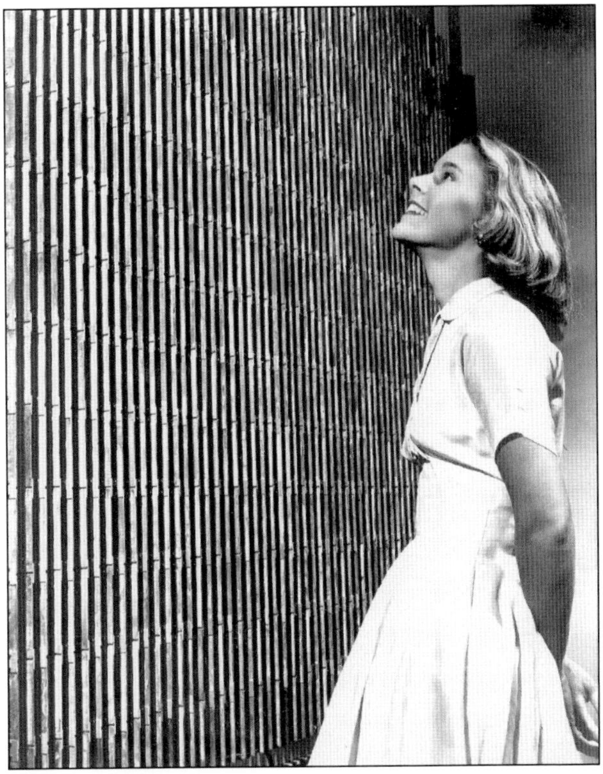

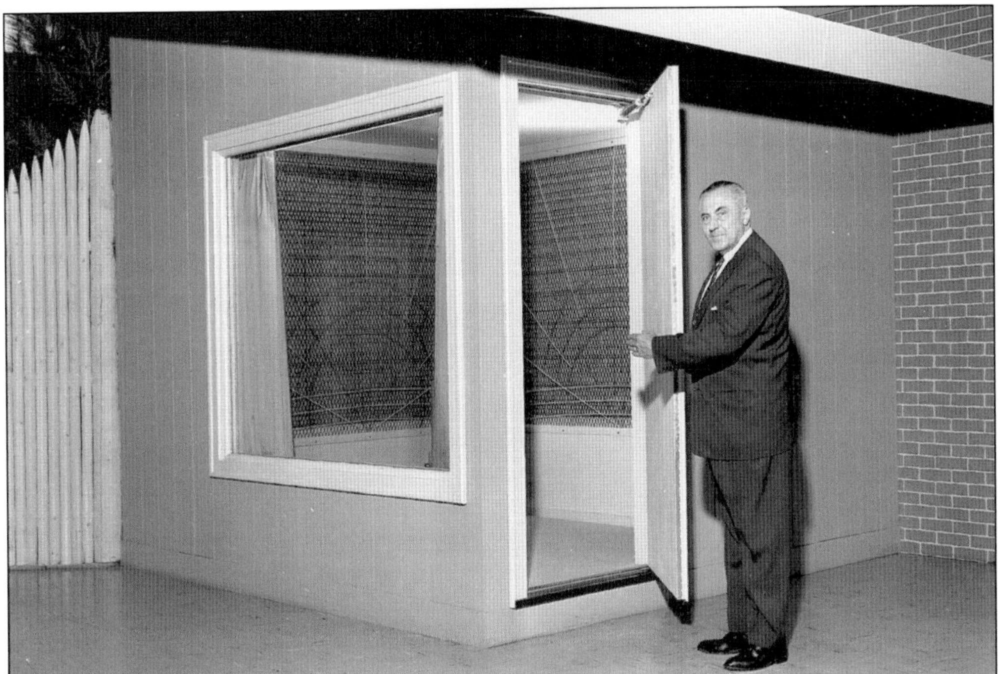

In 1956, Nils Lindenblad welcomes you to General Sarnoff's air-conditioned apartment, free of motor hum or drafts. The U.S. Navy contracted for RCA equipment for noiseless thermoelectric cooling for its submarine fleet, where silence is a virtue worth paying for above any other.

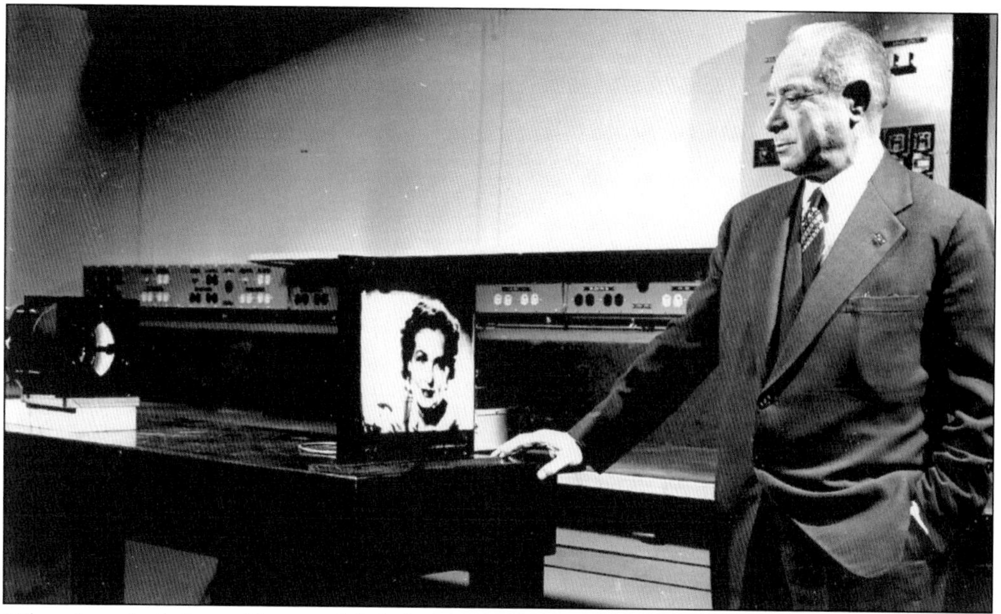

When he asked the staff to amplify light, Sarnoff was hoping for a flat-panel television, not a laser. He could not hide his disappointment with Benjamin Kazan's combination of charged photoconductive and electroluminescent panels, which displayed a still picture 1,000 times brighter than the projected input.

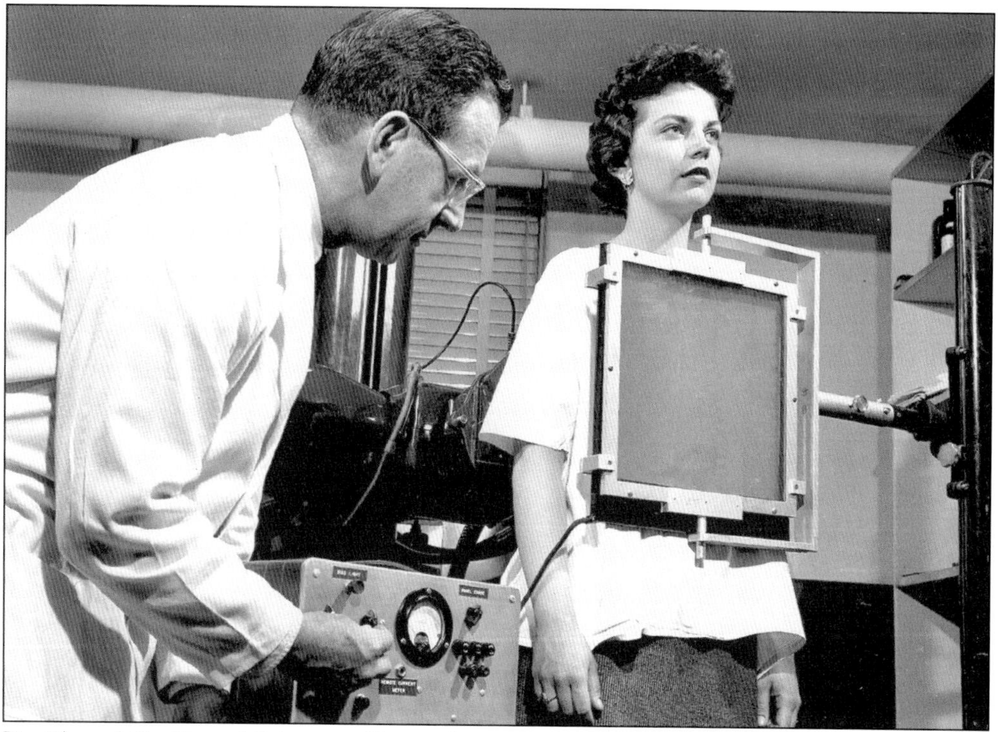

Dr. Elwood Godfrey of Princeton Hospital prepares to x-ray Beverly Must's chest as part of an evaluation of the center's x-ray amplifier to replace the traditional fluoroscopic screen in April 1958.

William D. Houghton looks over the magnetic tape used in the National Broadcasting Company's new color videotape recorder in May 1955. It was first used to record material transmitted from New York to St. Paul for the dedication ceremony of Minnesota Mining & Manufacturing Company's new research center.

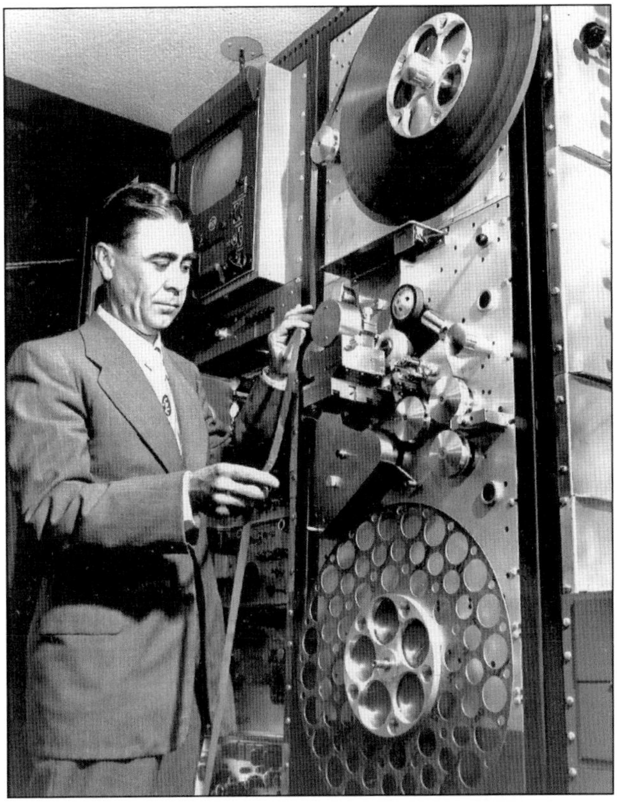

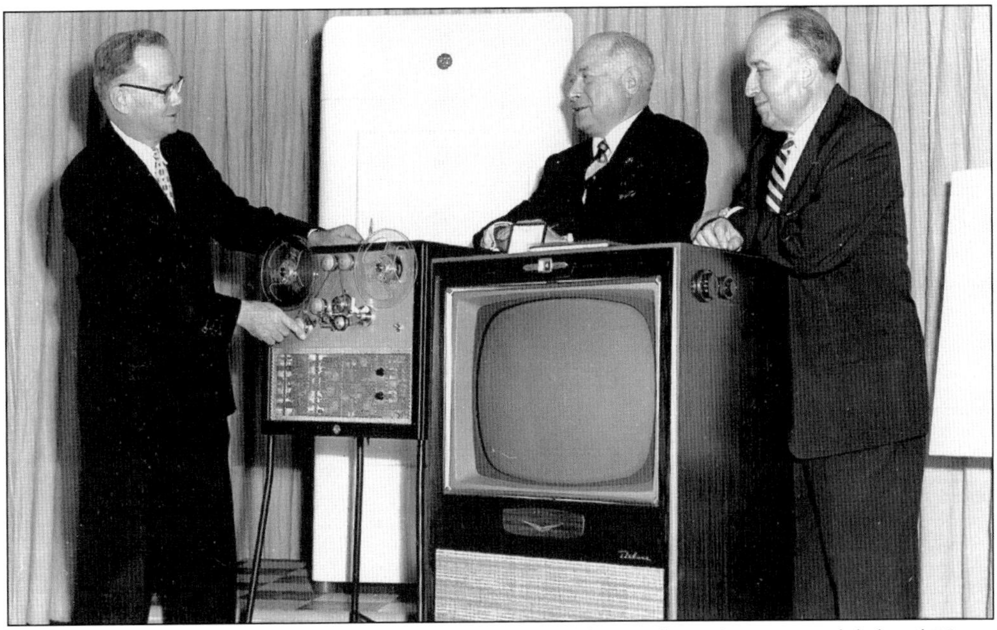

Center director Elmer Engstrom (left), David Sarnoff (center), and acoustics lab head Harry Olson look pleased with the home videoplayer, Sarnoff's third present. Olson's experience did not run far into magnetic recording; the result was that Ampex Corporation beat RCA to the practical broadcast videotape technology.

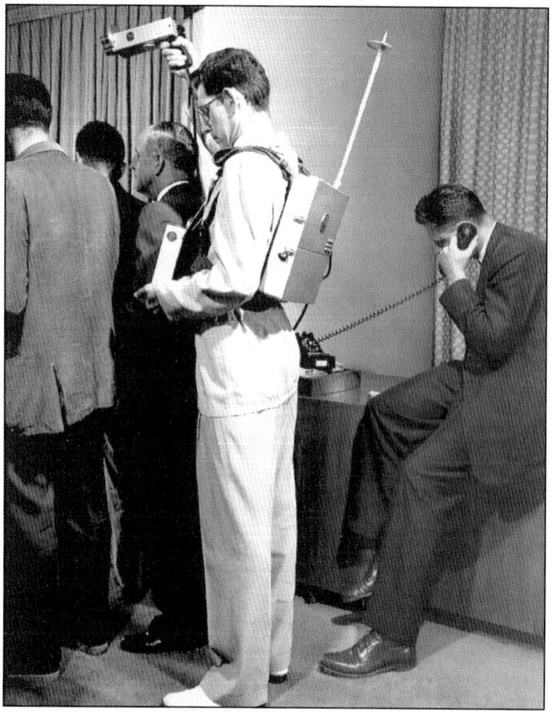

Wally Boyer, dapper in pincord and white bucks, looks into his viewfinder while his camera, the "creepy peepy," peers over Win Pike (left) and Les Flory. NBC used the system at the political parties' conventions that summer of 1956, and the Academy of Television Arts and Sciences nominated the RCA innovators for an Emmy. Best technical achievement, however, went to Ampex for its videotape recorder.

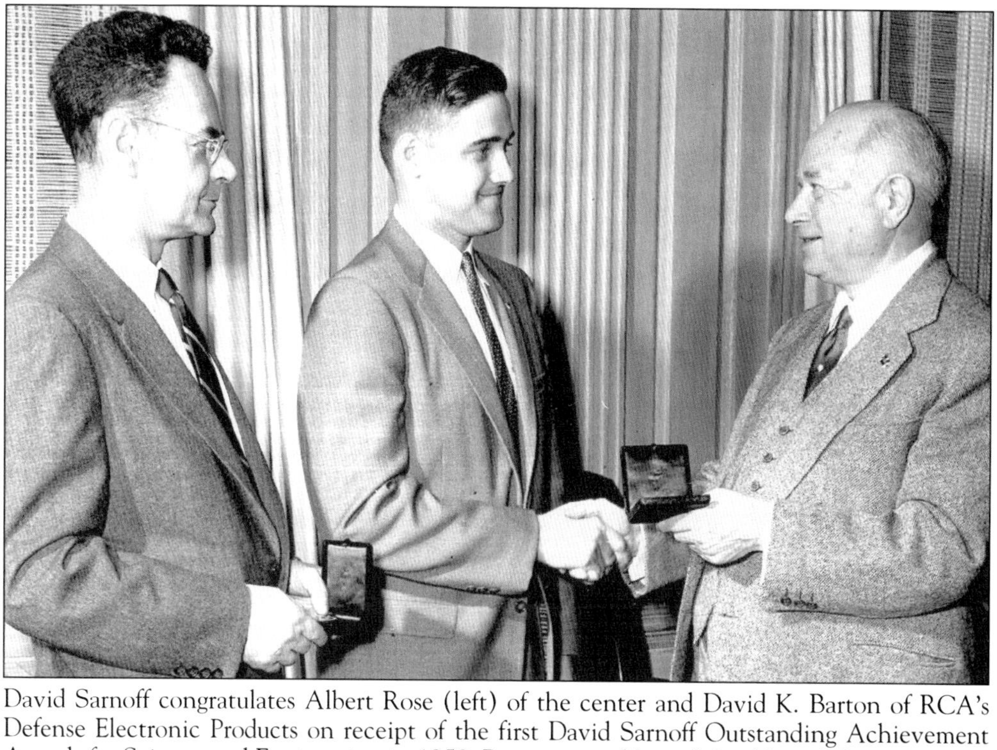

David Sarnoff congratulates Albert Rose (left) of the center and David K. Barton of RCA's Defense Electronic Products on receipt of the first David Sarnoff Outstanding Achievement Awards for Science and Engineering in 1958. Rose received his solid gold medallion "for basic contributions to the understanding and utilization of photoelectric phenomena" with regard to video camera tubes, and Barton his "for important contributions to precise tracking radars."

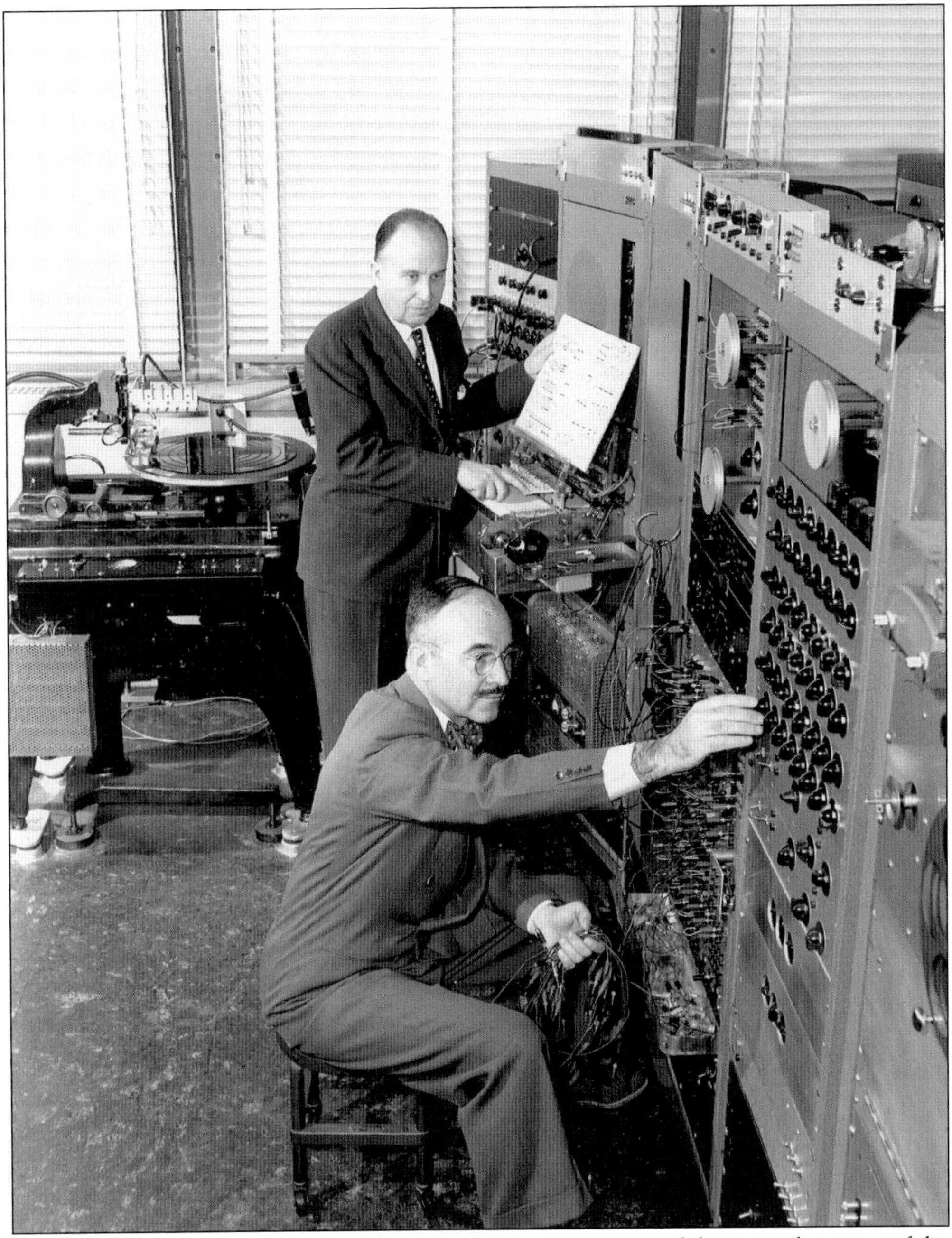

Harry Olson works the keyboard while Herbert Belar adjusts one of the controls on one of the world's first electronic music synthesizers. Developed with the intention of replacing orchestras for popular mood music and soundtracks, a later model had the ironic fate of donation to the music department at Columbia University for use in experimental compositions.

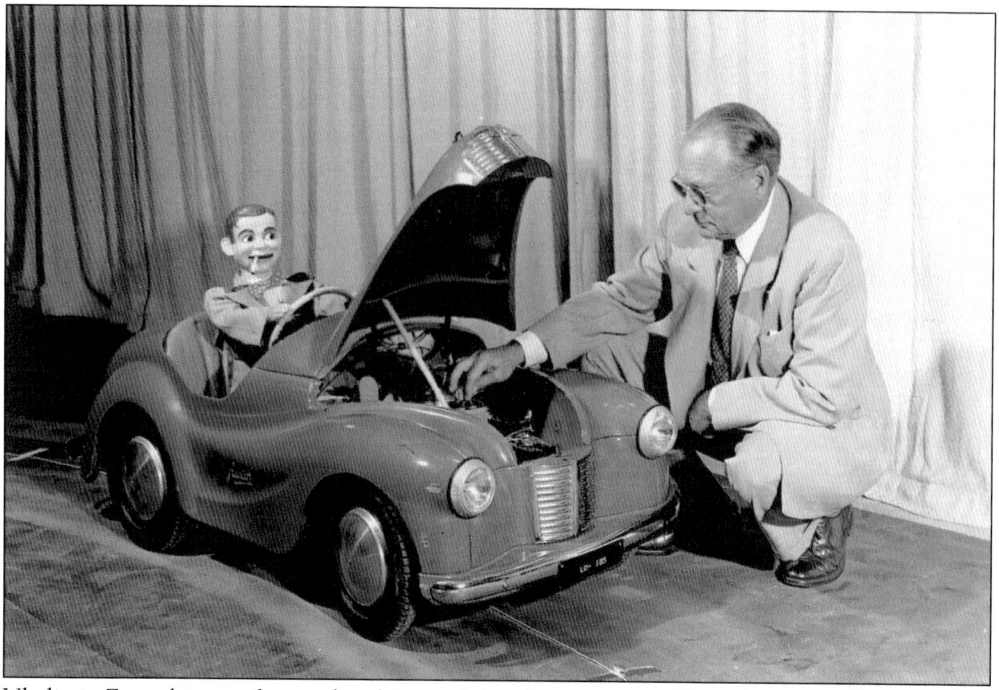

Vladimir Zworykin confronts the driver of the first version of his electronically controlled highway in the center's basement in the early 1950s. This marked the beginning of the center's long association with the automotive electronics industry.

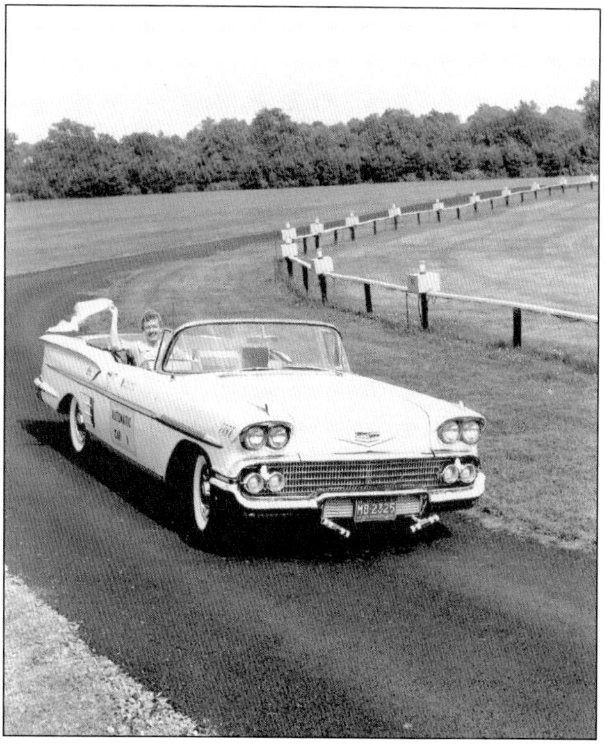

Guided by radio signals from embedded sensors in the road and the fence, the woman in the car can throw her hands up in the air and leave the driving to the ether in 1958.

In this 1950s view, an experimental facsimile printout provides food for thought.

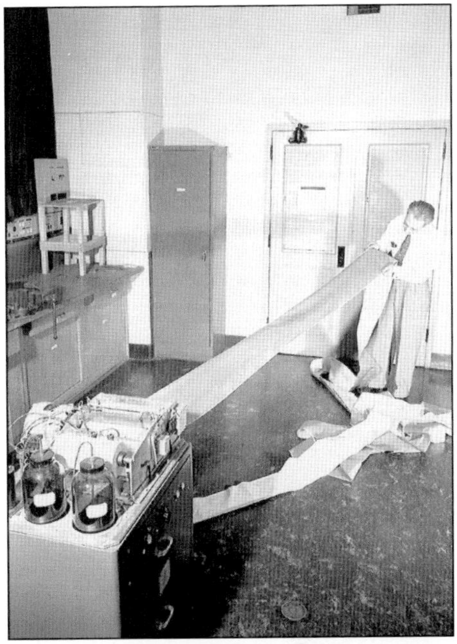

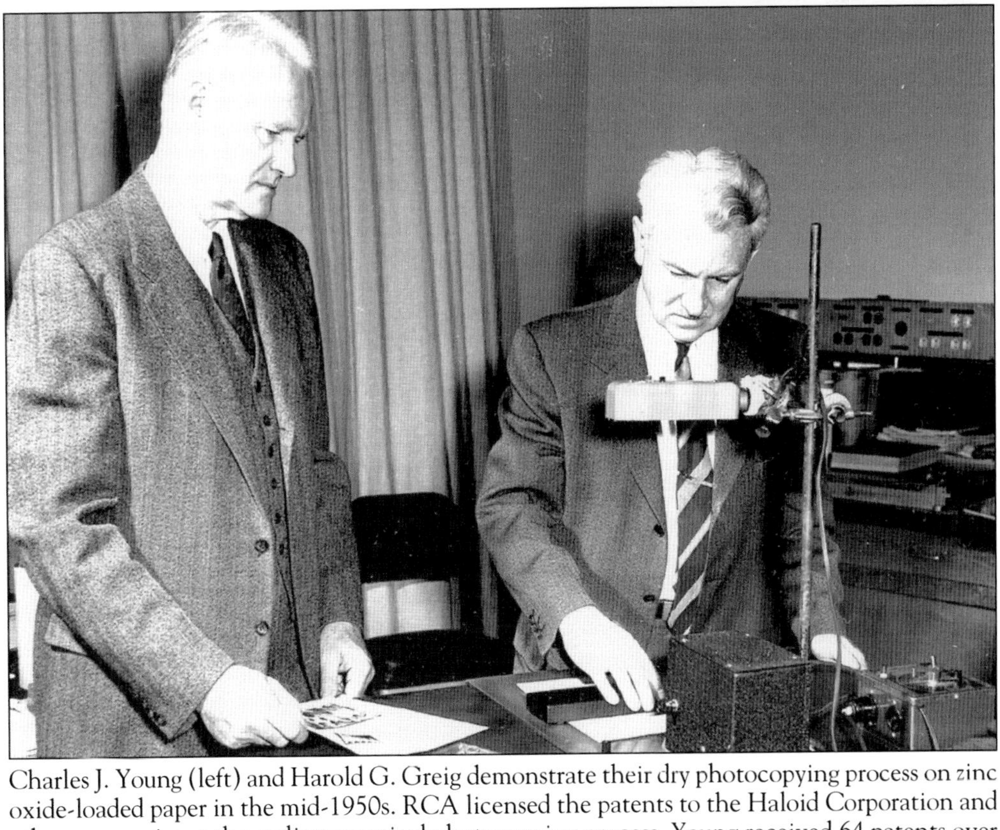

Charles J. Young (left) and Harold G. Greig demonstrate their dry photocopying process on zinc oxide-loaded paper in the mid-1950s. RCA licensed the patents to the Haloid Corporation and other companies as the earliest practical photocopying process. Young received 64 patents over a career in facsimile innovation that began in 1922; Greig coauthored with Chester Carlson of Xerox fame the article on electrophotography in the classic work on the subject.

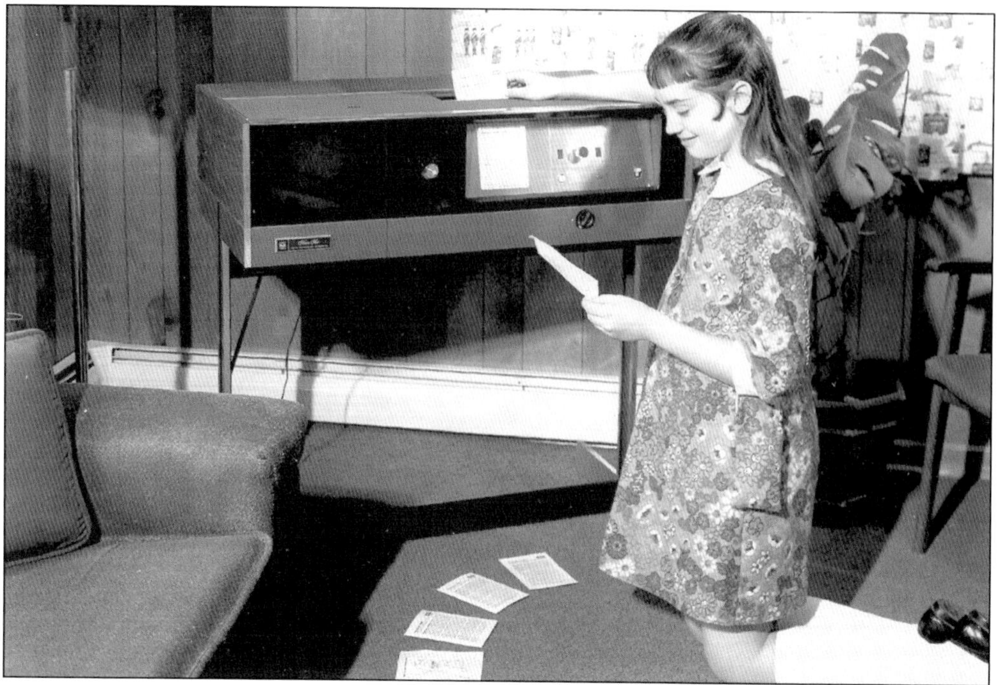

Ten years after Electrofax and 15 after Ultrafax, the center's innovators debuted Homefax in 1968. Connected to the rising interest in home information systems, Homefax provided all members of the family with information they needed, from weather reports to recipes to homework assignments.

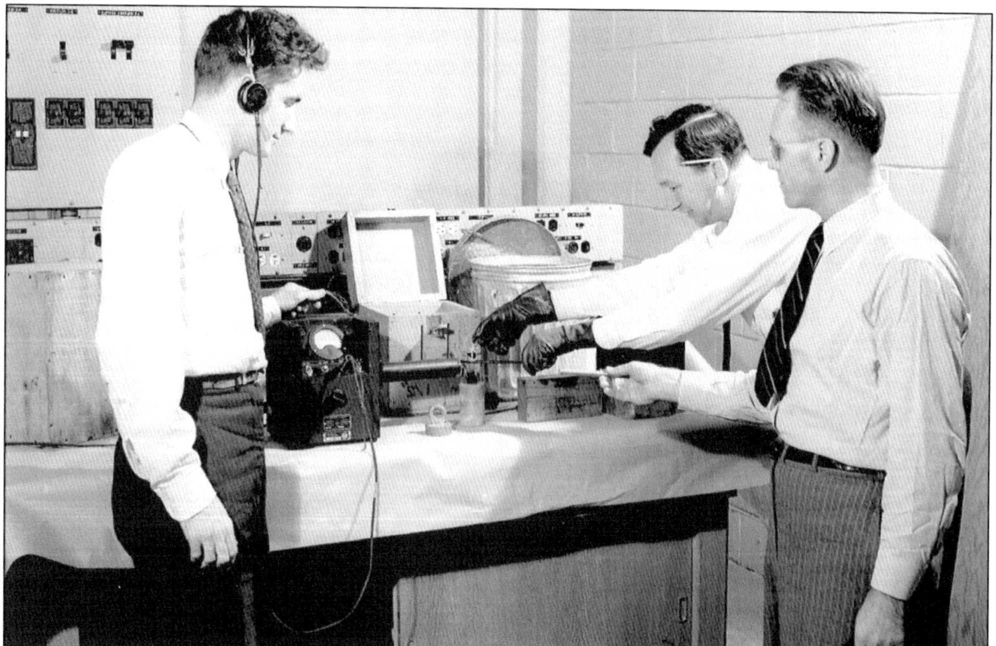

Schuyler Christian (center) and Ernest Linder (right) remove a radioactive isotope very carefully from its shipping container as technician J. Rogers monitors the Geiger counter sometime in the 1950s.

Paul Rappaport, father of modern solar cell technology, prepares to insert a bit of strontium-90 into the hole at top left in January 1954. Radiation from the strontium-90 triggers electronic amplification on the order of 200,000 to 1 in a semiconductor wafer and transistorlike junction, generating a tone from the earphone.

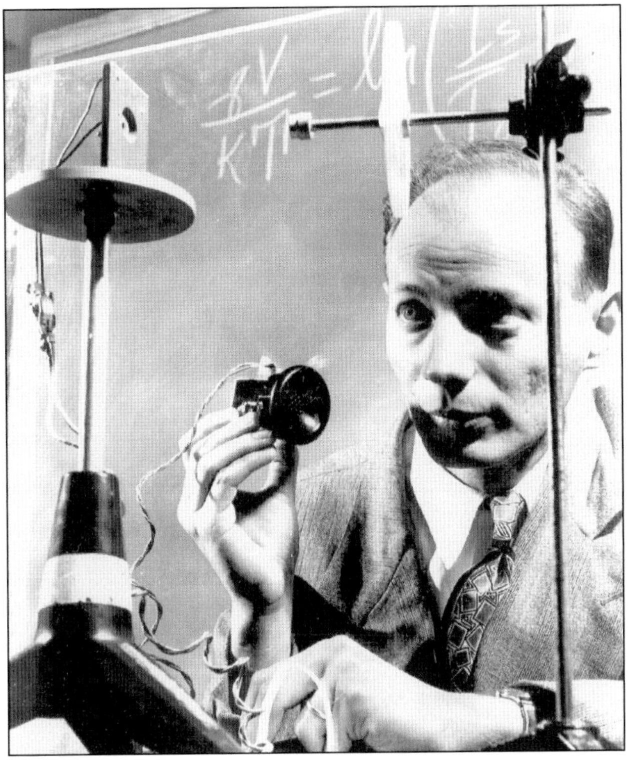

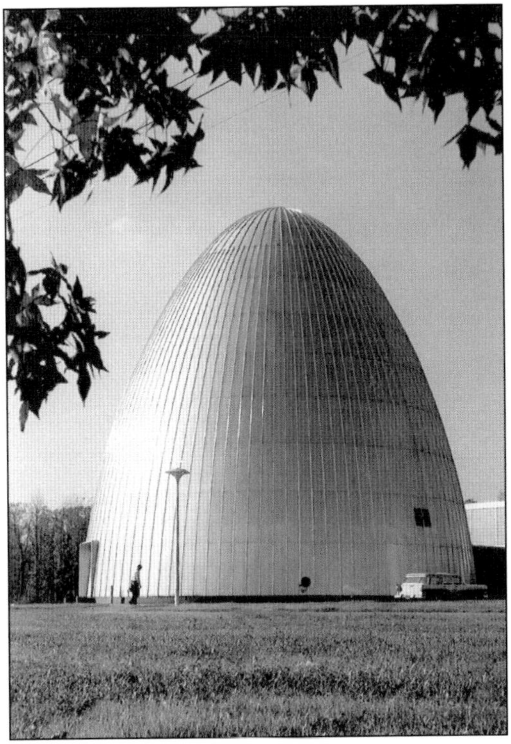

To get the radioactive isotope, RCA was part of an industrial consortium that contributed to the construction and maintenance of the Industrial Reactor Laboratory off Schalks Crossing Road in Plainsboro. Privately owned by a nonnuclear company today, the building (designed by Skidmore, Owings & Merrill) is visible from parts of the adjacent Audubon Bird Preserve.

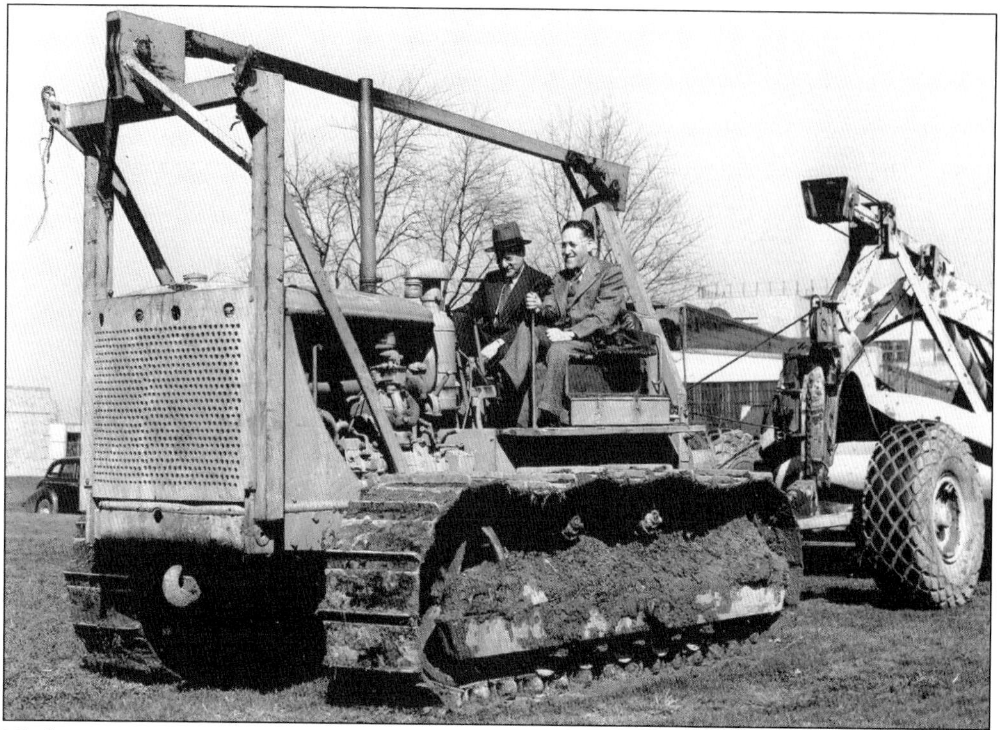

Vladimir Zworykin (left) and Irving Wolff take a bulldozer out for a spin during one of the frequent episodes of construction at the center in the early 1950s.

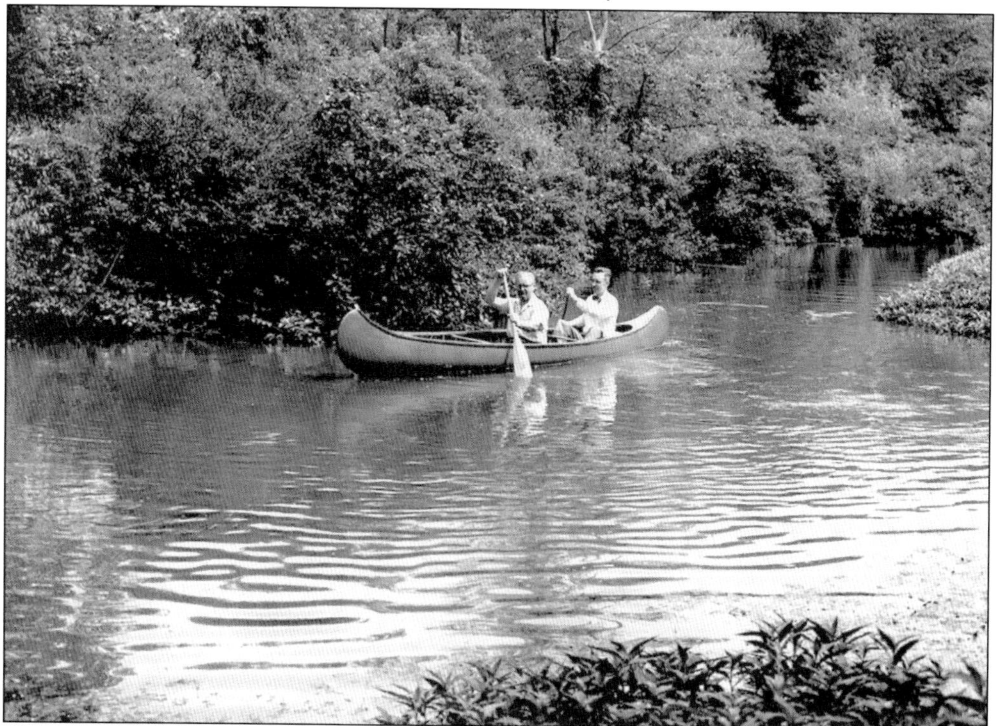

Two members of the technical staff took to canoeing to work via the Millstone River.

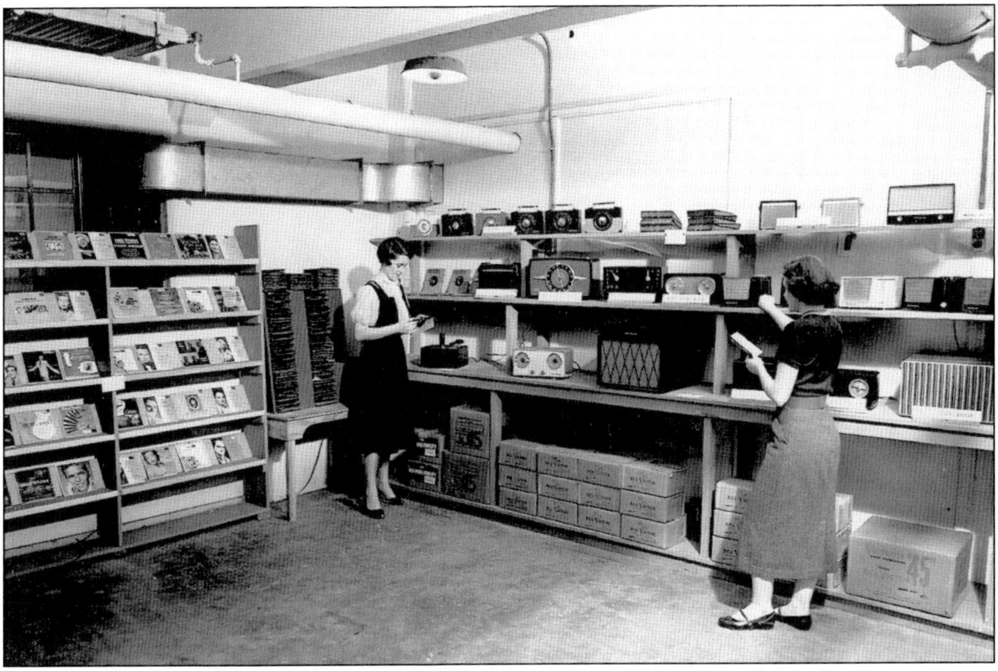

The RCA company store in the basement, pictured *c.* 1953, provided for all of one's home entertainment and electronic needs via mail order from RCA Victor in Camden.

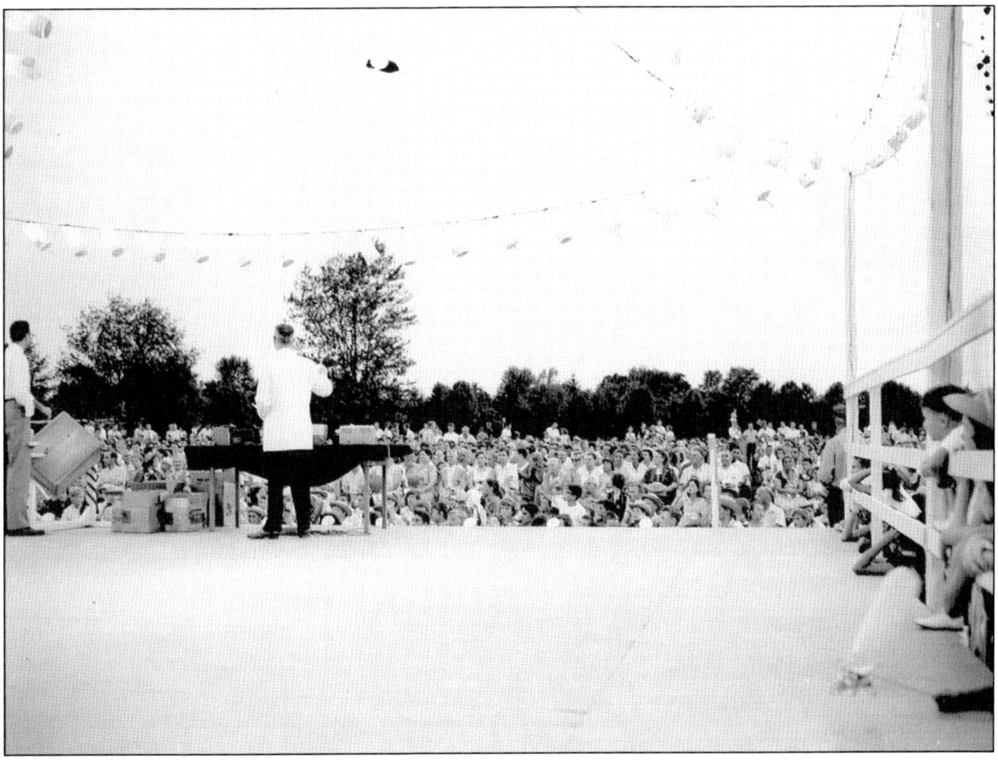

At the 1962 picnic, RCA consumer electronics are raffled off on the front lawn.

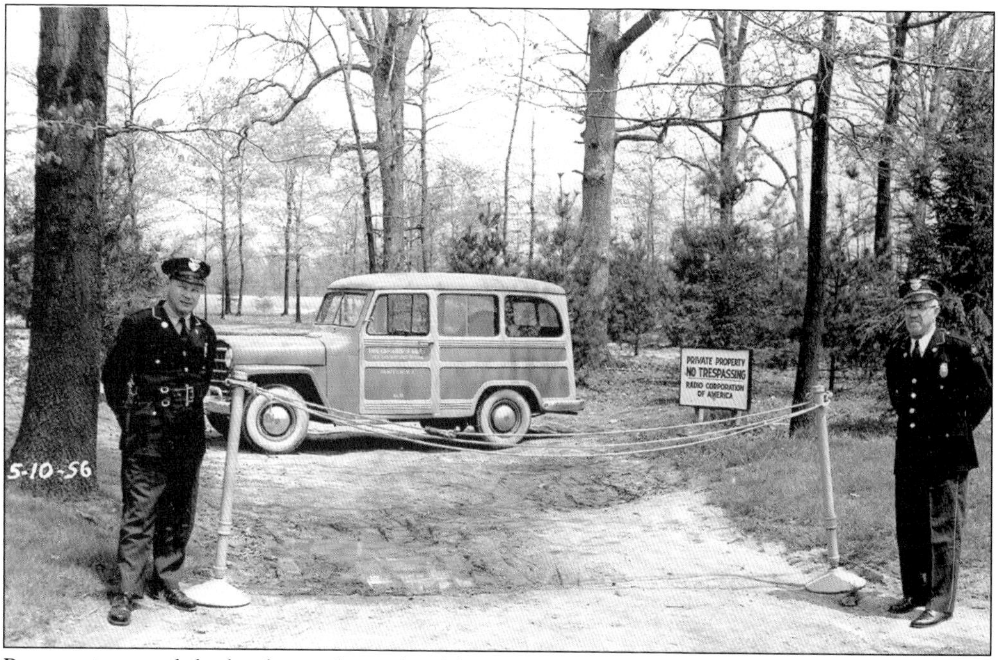

Because it owned the land on either side of the right of way to the Millstone River Sheep Wash, RCA had to assert its ownership of the way by closing it to public travel for 24 hours. This took place in May every year until at least 1964.

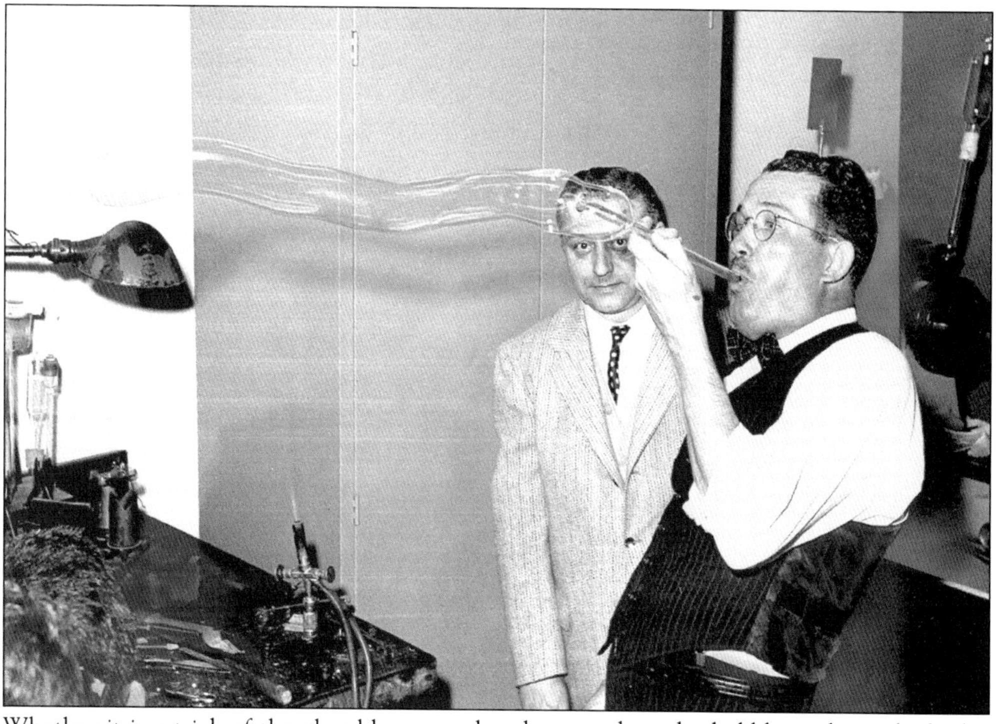

Whether it is a trick of the glass blower or the photographer, the bubble at the end of John McCool's pipe is an impressive one. Glassblowing was an essential craft for electron tube research and development.

Herbert Kroemer spent "three of the happiest years" of his life at the center between 1954 and 1957. During that time, he made the drift transistor commercially feasible and established the prospects and problems involved in using wide bandgap heterogeneous semiconductors. Examples include light-emitting diodes, lasers used in CD drives, and the chips that power cell phones. He shared the Nobel Prize in physics in 2000 for his career of accomplishment in his chosen and globally important field.

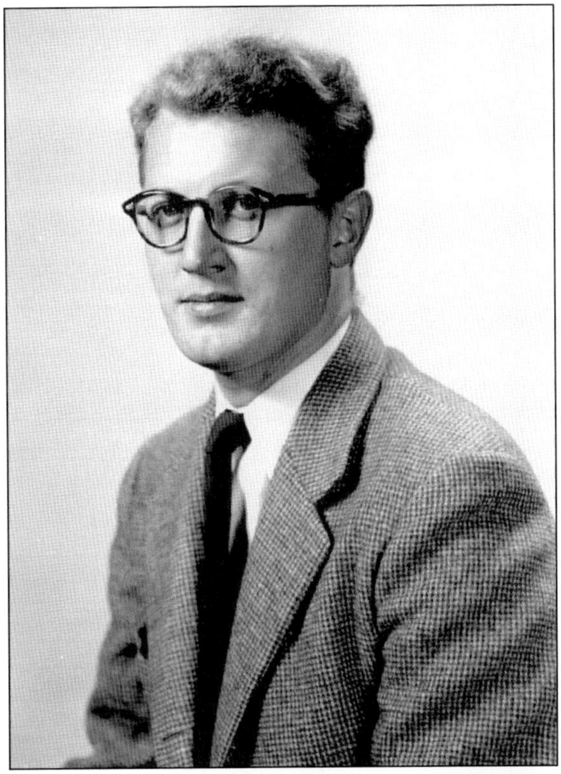

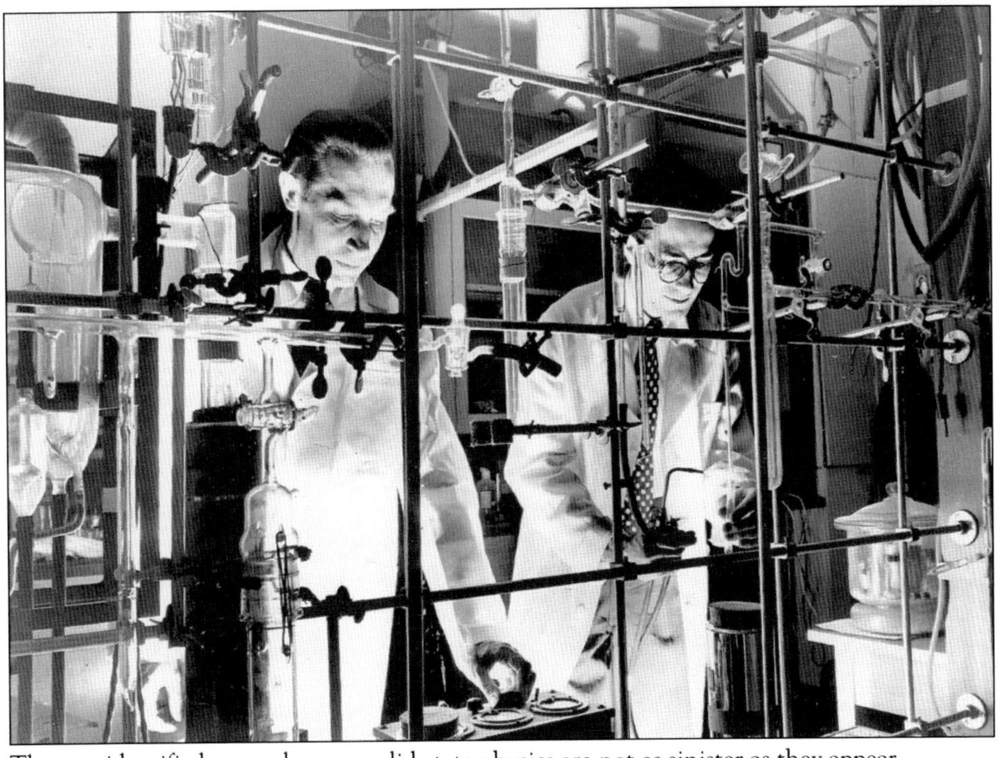

These unidentified researchers on solid-state physics are not as sinister as they appear.

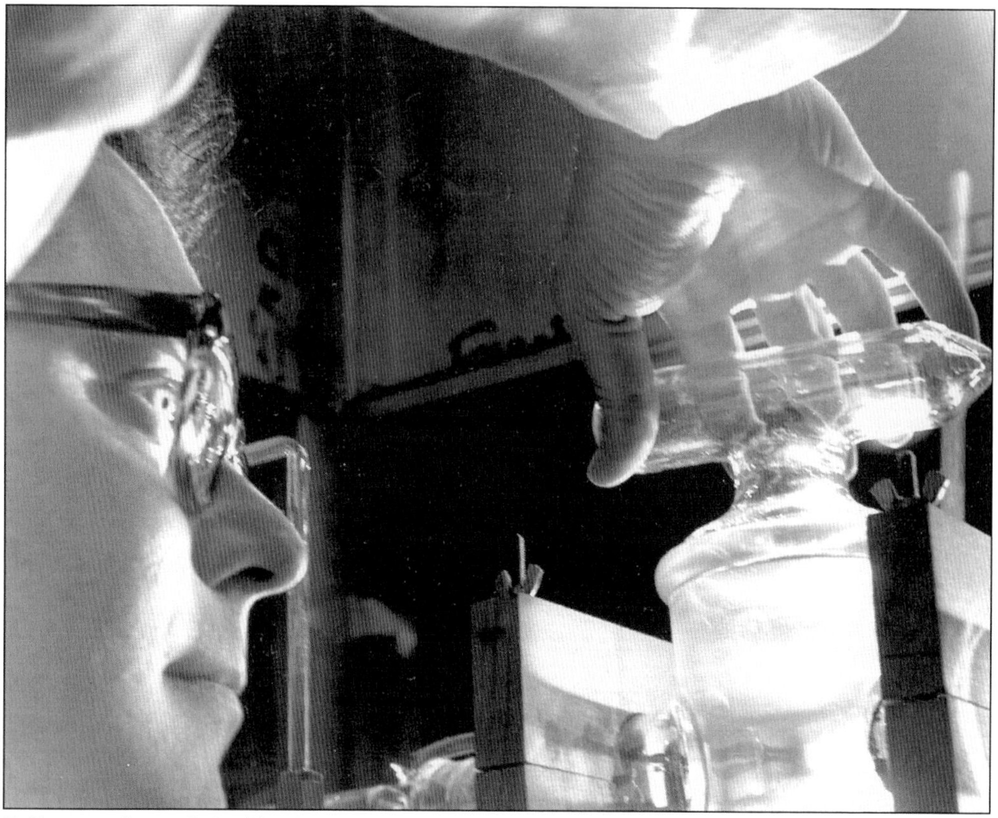

Following the path laid by Kroemer, in 1966 James Tietjen opens the stopcock to begin gas flow in the vapor phase epitaxy (VPE) process to grow crystals of gallium arsenide and gallium phosphide. The process, invented by Tietjen and Herbert Nelson, marked the beginning of all the epitaxial procedures used routinely today in semiconductor fabrication.

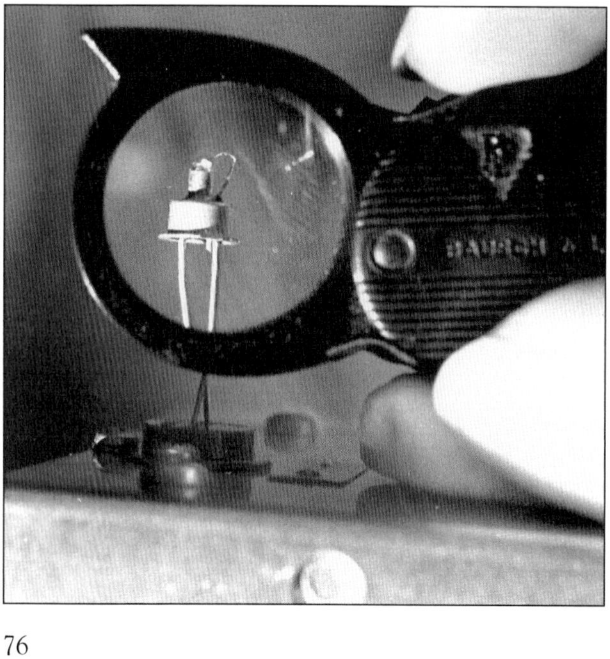

This light-emitting diode penlight, grown by VPE, is one consequence of Kroemer's revolution in semiconductors.

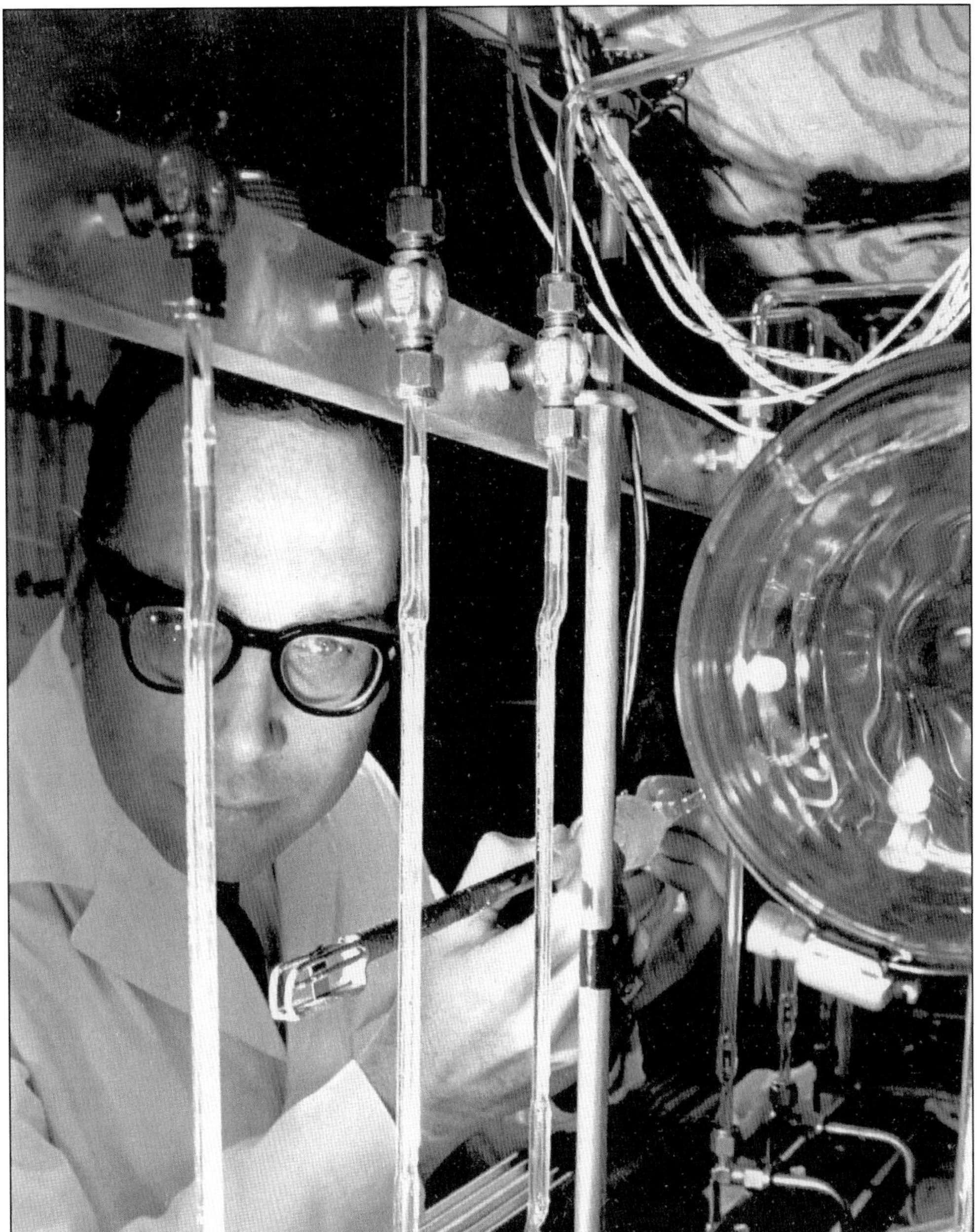

Ronald Enstrom checks a grown wafer of gallium arsenide standing vertically in the cagelike extension at the end of a quartz rod. The wafer was grown by VPE through controlled deposition from a mixture of hot gases.

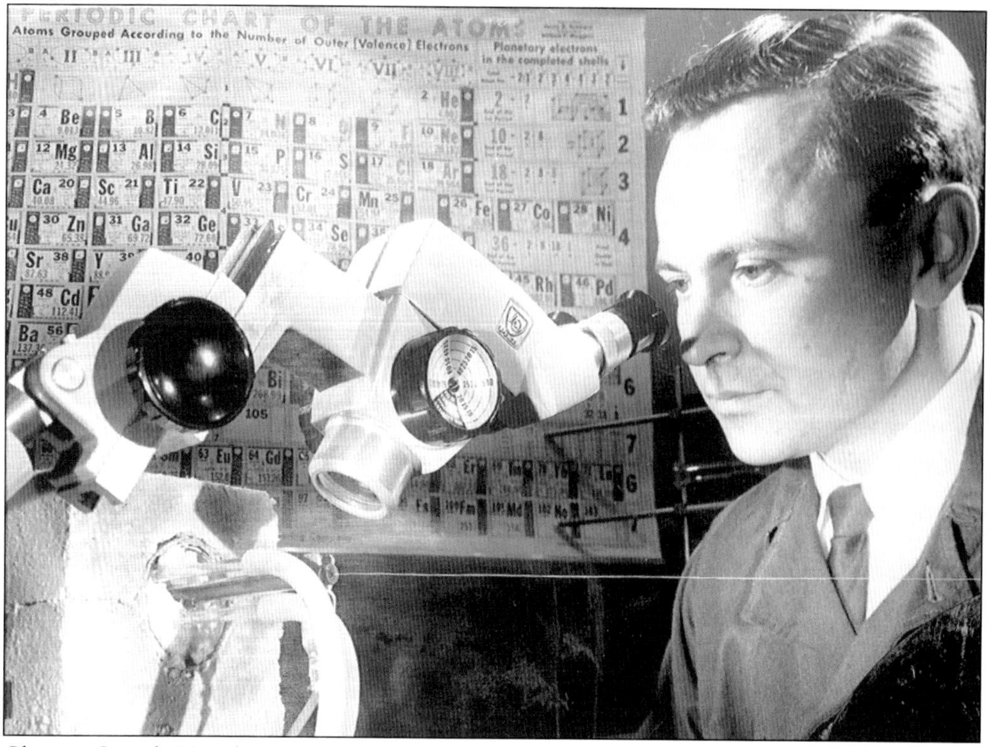

Chemist Joseph Hanak analyzes the application of chemical vapor deposition techniques to the manufacture of niobium-tin alloy in 1961, which led to the center's world leadership in superconducting magnets.

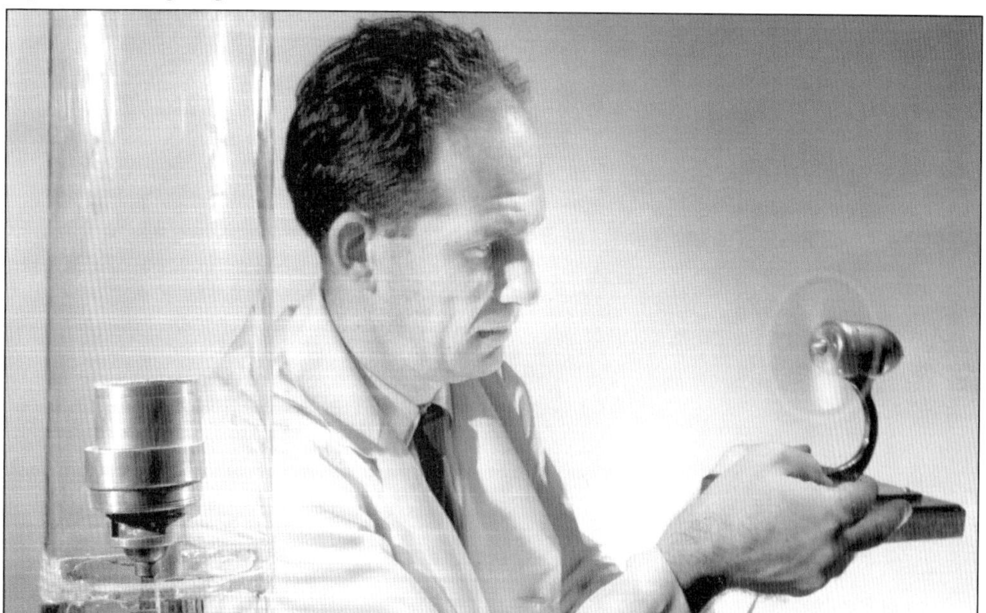

Ben Abeles stays cool with a fan electrically powered via the material research team's new alloy of silicon-germanium in 1963. The alloy converts heat to electricity very efficiently and later made possible the deep-space travels of the Voyager, Galileo, and Cassini spacecraft.

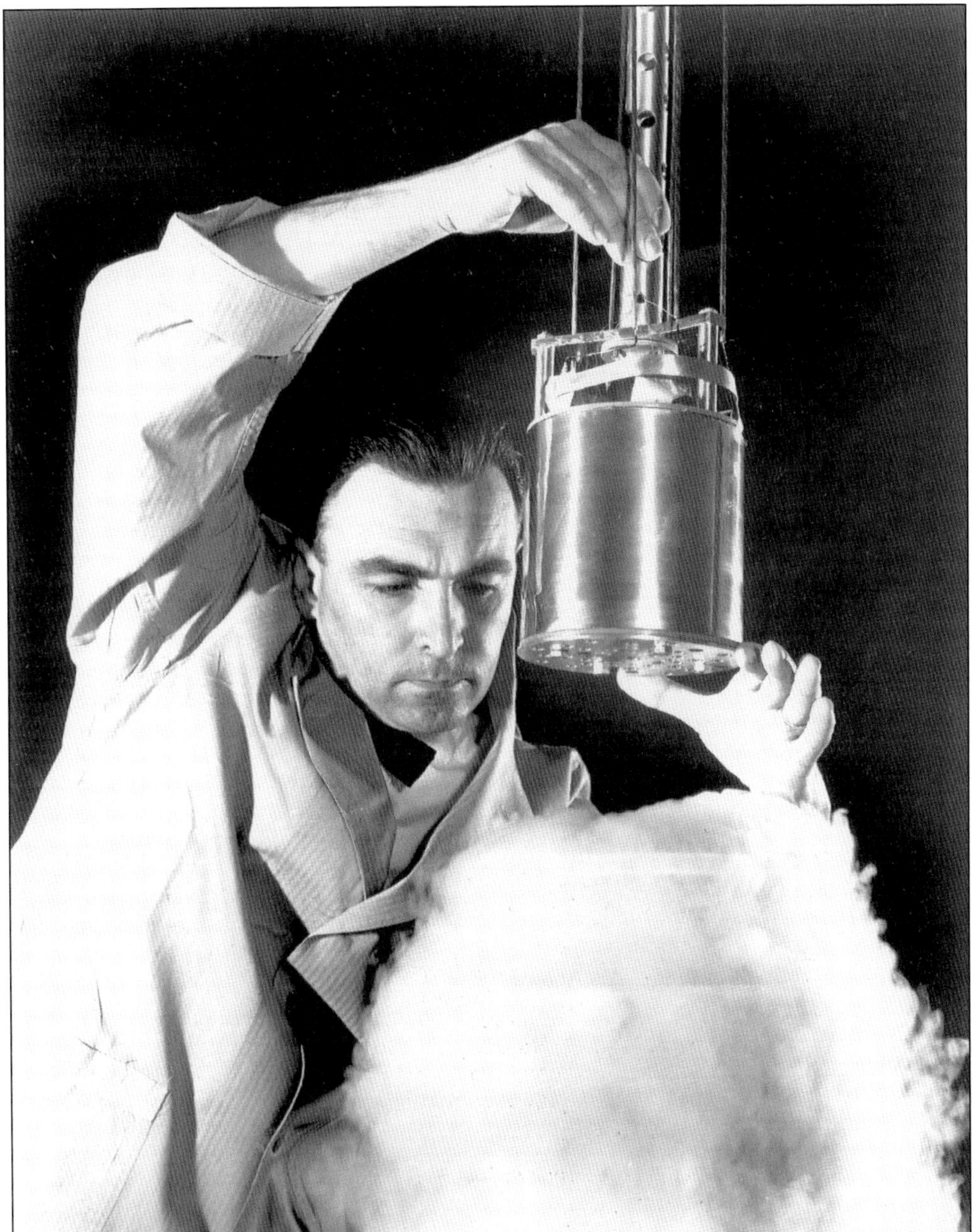

Peter Kolondra prepares to lower a magnet into its bath of liquid helium in 1964, enabling the world's most powerful superconducting magnet to generate a field of 10.7 Tesla. Led by Fred Rosi, the staff on this project worked with RCA Harrison to commercialize these magnets within two years.

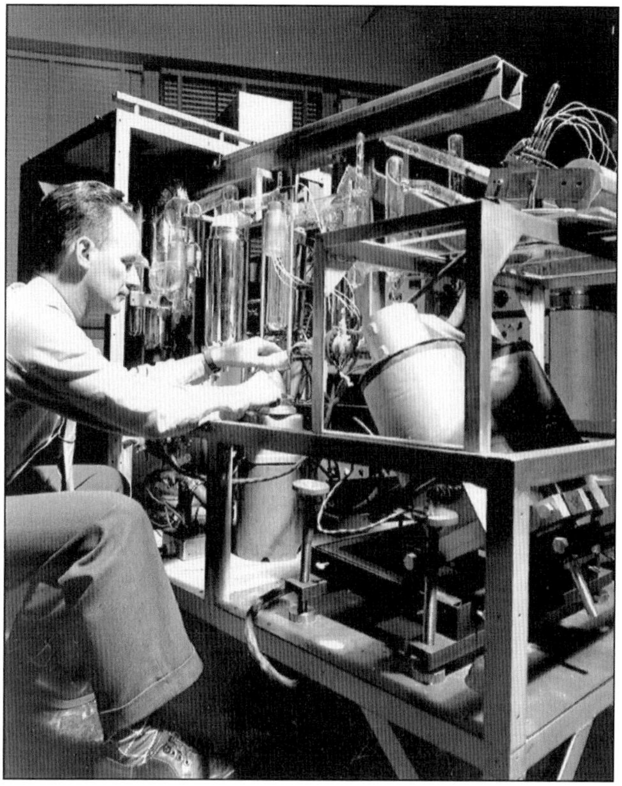

To make these technologies work at a subatomic level, one needs very large, complex, and expensive equipment, like the mass spectrometer being prepared here for a trial run.

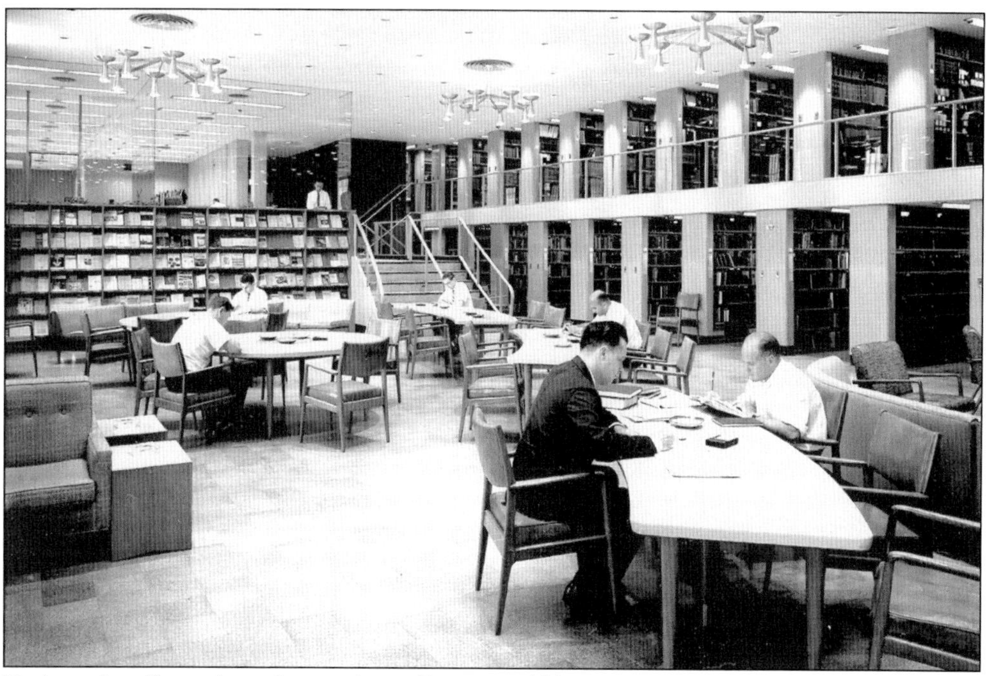

Technical staff members also need a well-equipped library as a source of specific, current, and relevant information. Librarian Fern Cloak made sure the reading room stayed quiet even without carpeting, but smoking was permitted.

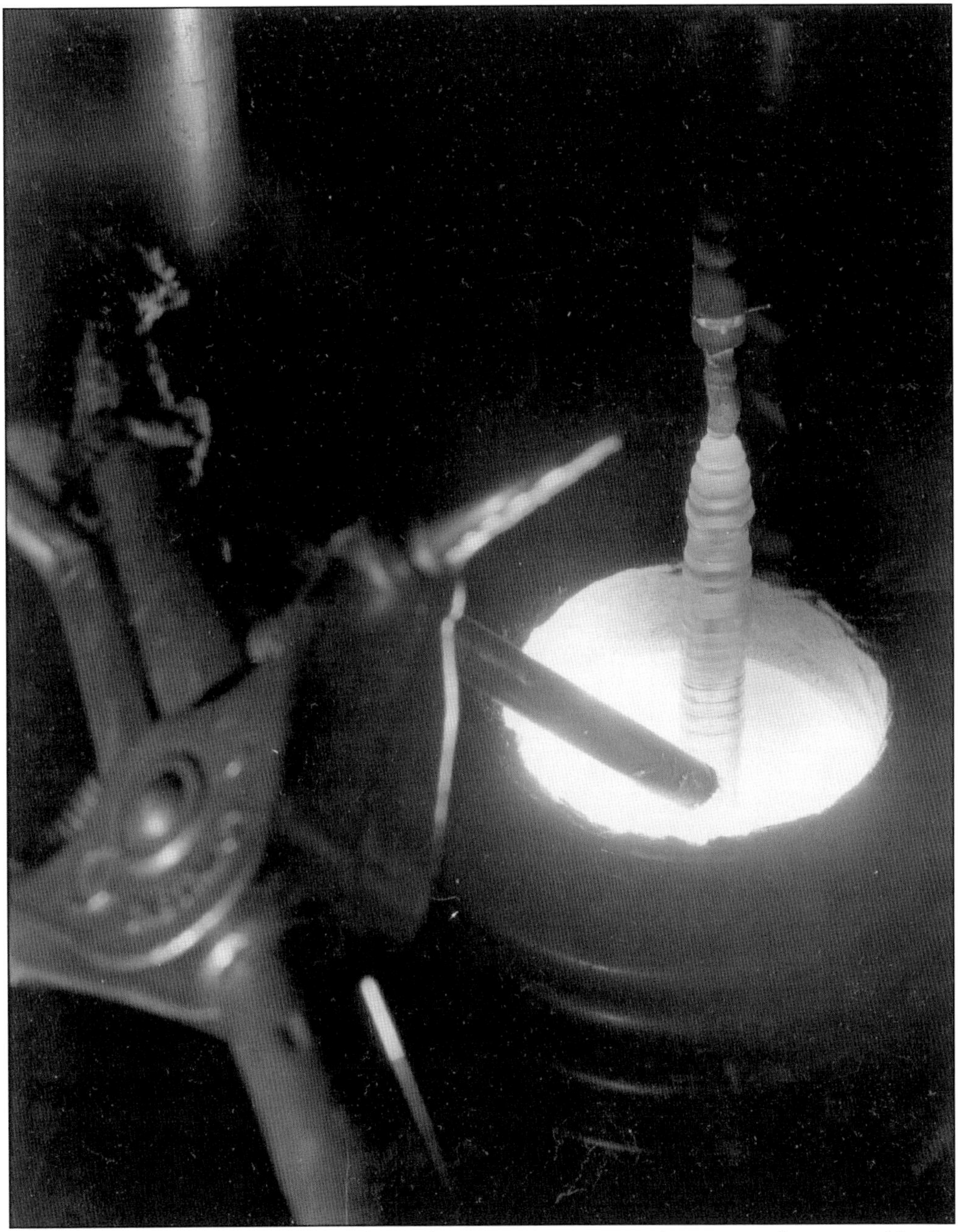

Pictured here is the formation of a calcium tungstate crystal in an iridium crucible for experimentation as a material for lasers in the early 1960s.

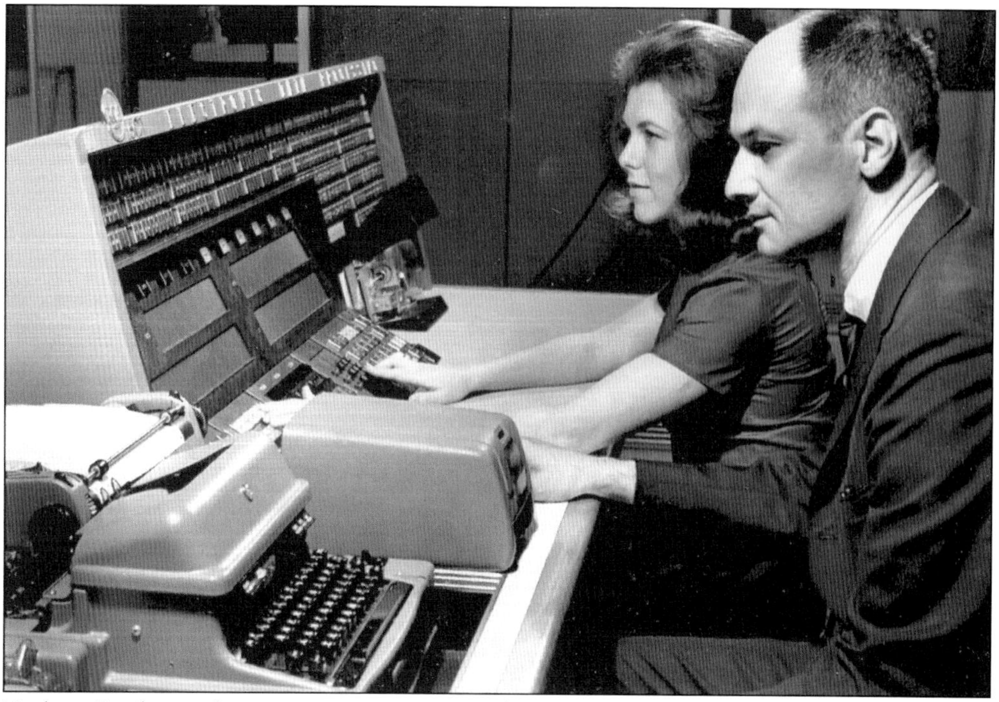

Nathan Gordon and an associate operate another large and useful piece of equipment—the RCA 601 computer in the new systems research center at the end of the east wing *c.* 1963.

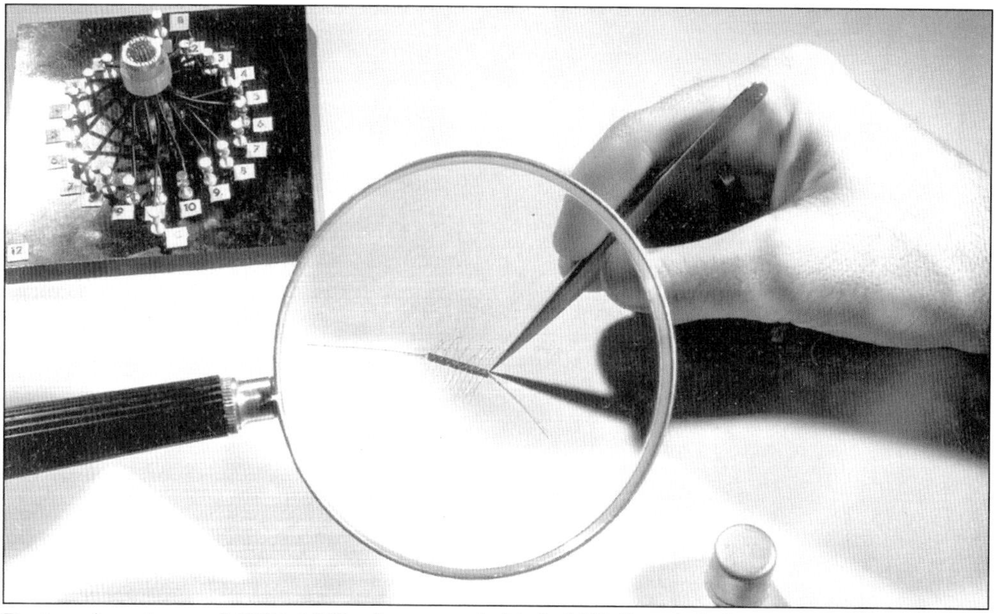

During the winter of 1957–1958, the center's Torkel Wallmark led a contract for the U.S. Air Force to develop an integrated electronic shift register. RCA prematurely announced Wallmark's integrated circuit and predicted that it would make possible the packaging of 100 million elements into a cubic foot. Wallmark, however, could not quite make his device work. That summer, Jack Kilby at Texas Instruments and Robert Noyce at Fairchild Semiconductor independently announced working integrated circuits.

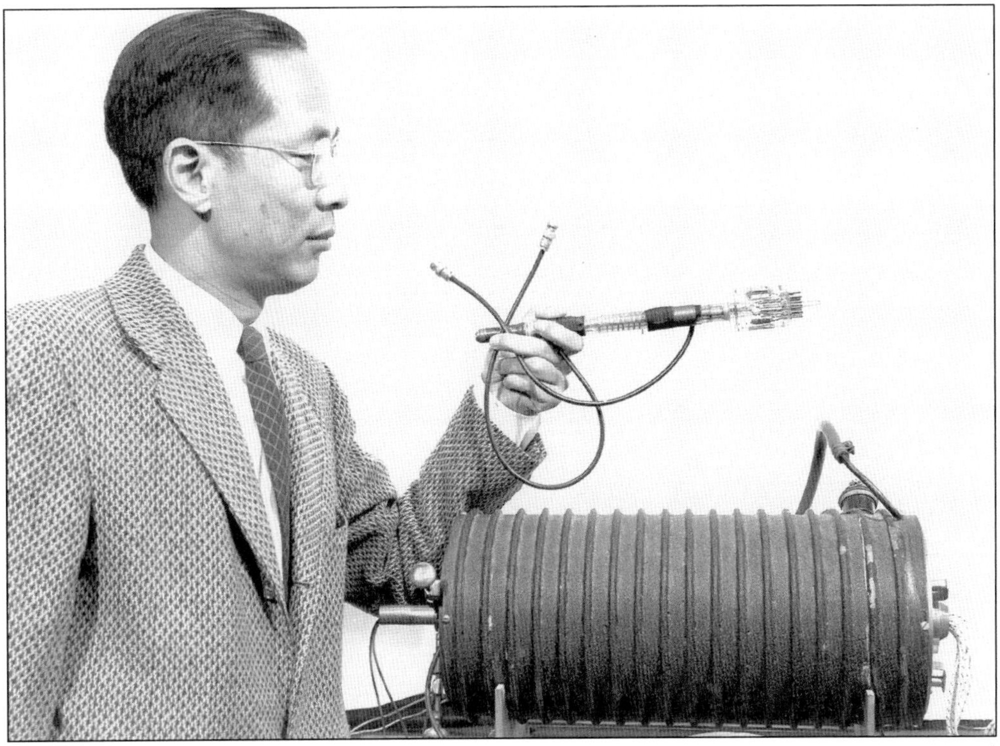

Kern Chang holds a lightweight traveling wave tube for high-power microwave signal amplification in 1958.

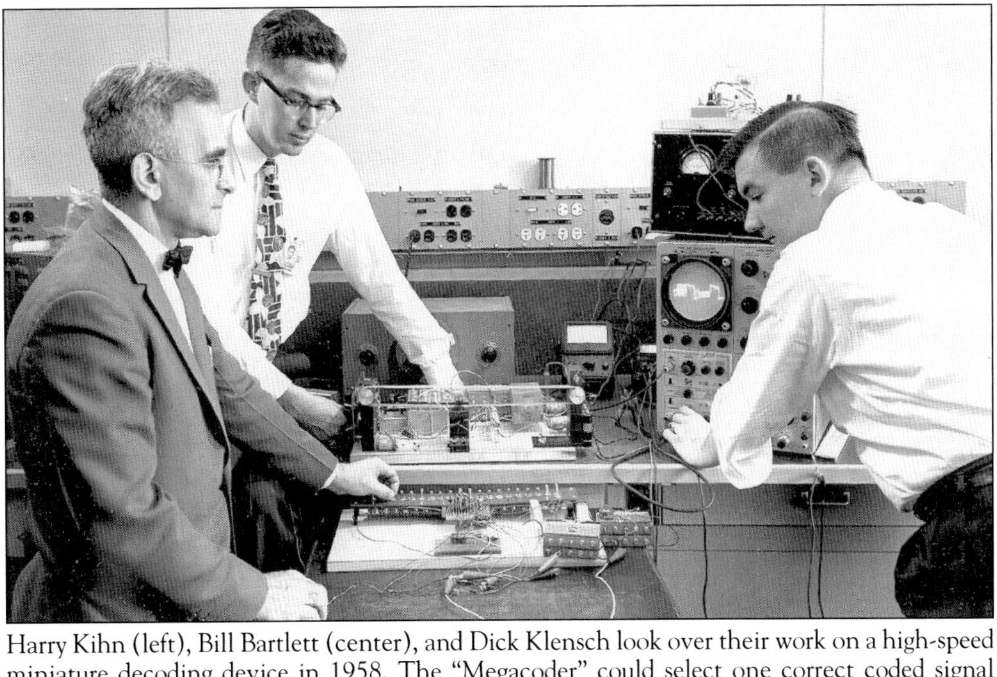

Harry Kihn (left), Bill Bartlett (center), and Dick Klensch look over their work on a high-speed miniature decoding device in 1958. The "Megacoder" could select one correct coded signal out of a million transmitted from a central location and represented one approach to cell phone technology.

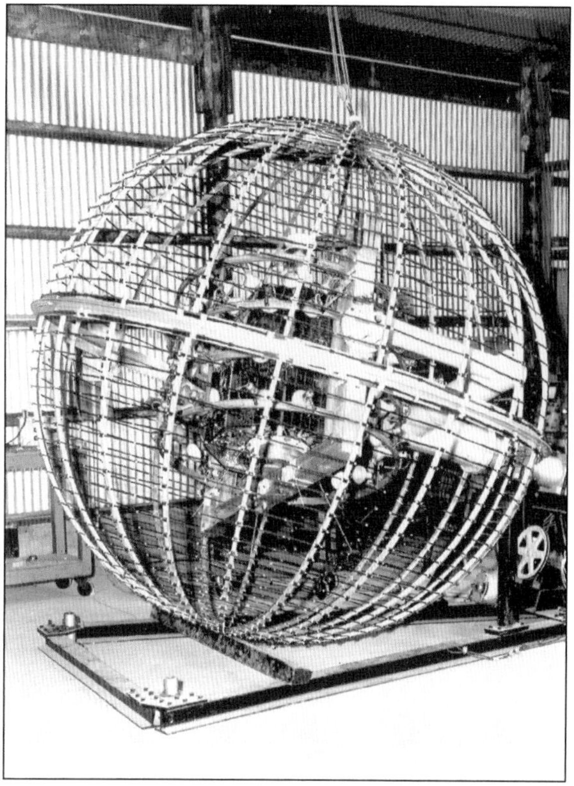

The Relay satellites of 1962–1964 were RCA's first for communications and provided the first transatlantic television transmissions in 1963. The center initiated satellite research in 1951, leading to the establishment of RCA's Astro-Electronics Products in Hightstown, New Jersey, in 1958, just after the federal government formed the National Aeronautics and Space Administration (NASA).

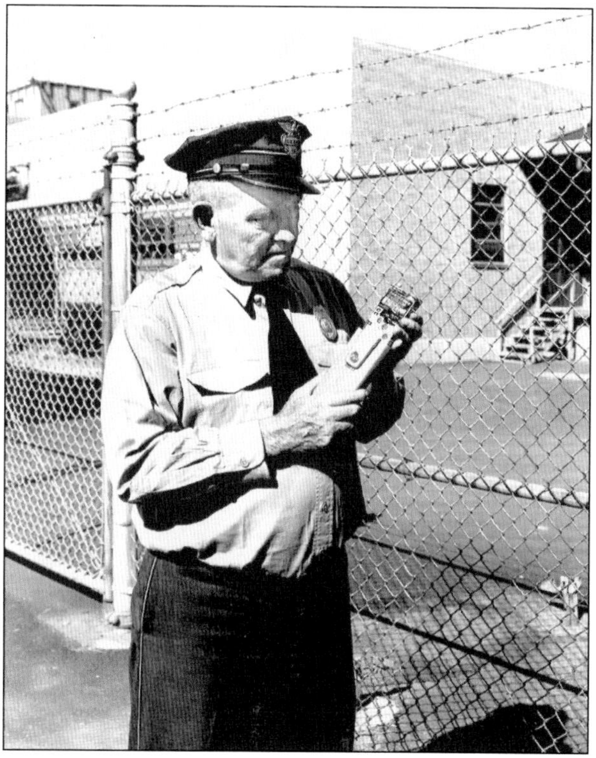

One of the center's security guards considers using his Megacoder.

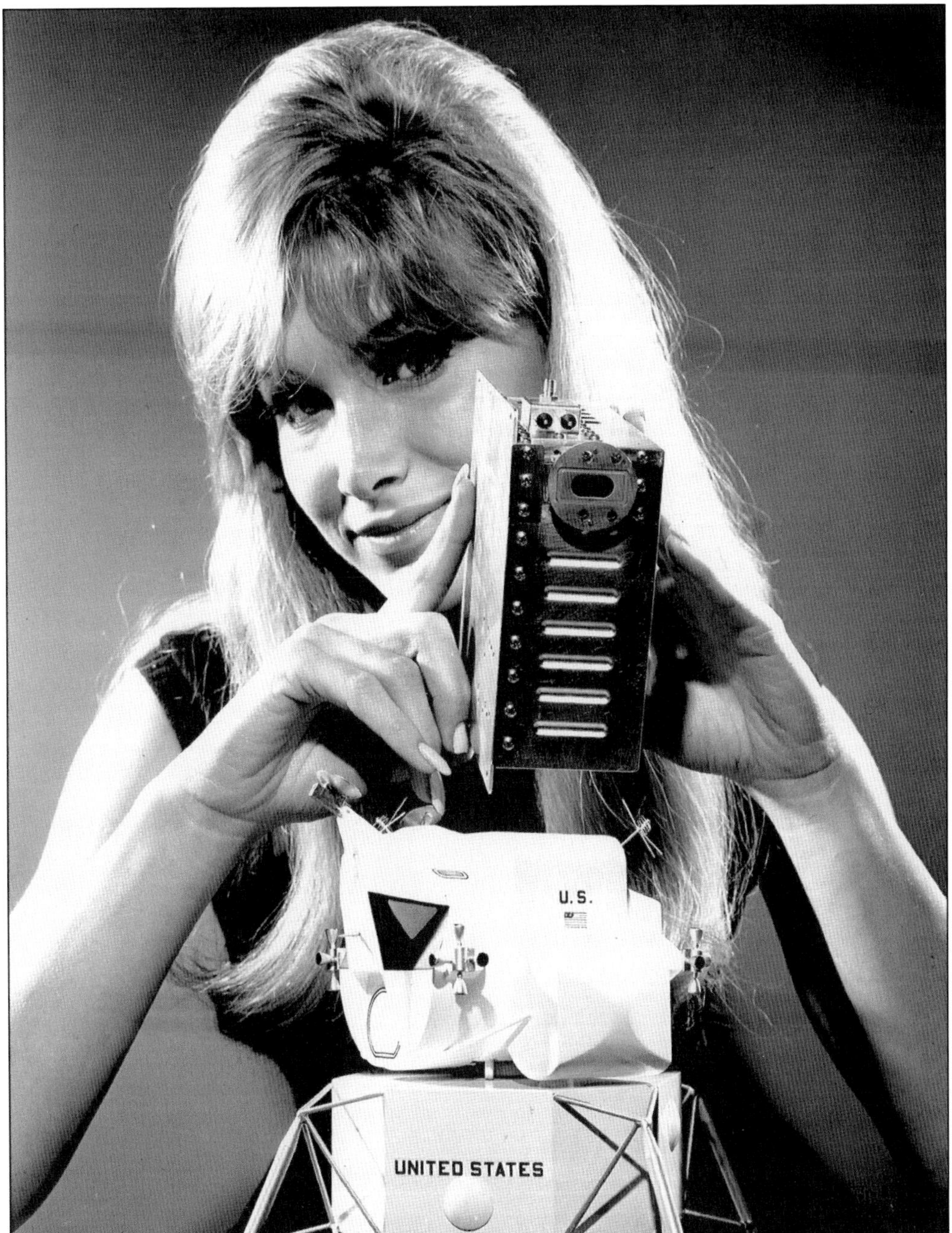

The Microwave Advanced Research Laboratory at the center was responsible for much of the communications hardware on the Apollo expeditions between the Lunar Explorer Module and the Lunar Orbiter, including this frequency multiplier.

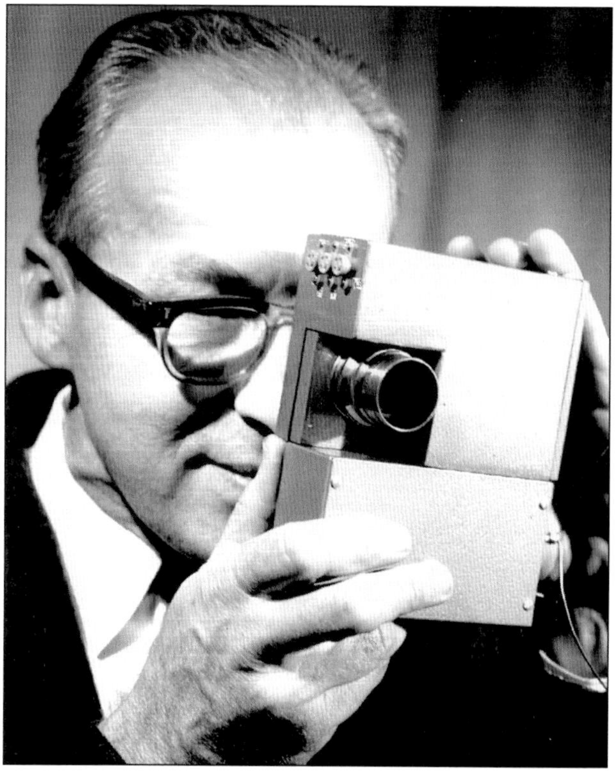

Paul Weimer devoted his career to improving video camera technology, starting with the image orthicon and his invention of the vidicon tube. After retraining in solid-state physics at the Sorbonne at RCA's expense, he invented the first working thin-film transistor in 1961, which he and four associates turned into the first solid-state camera in 1964.

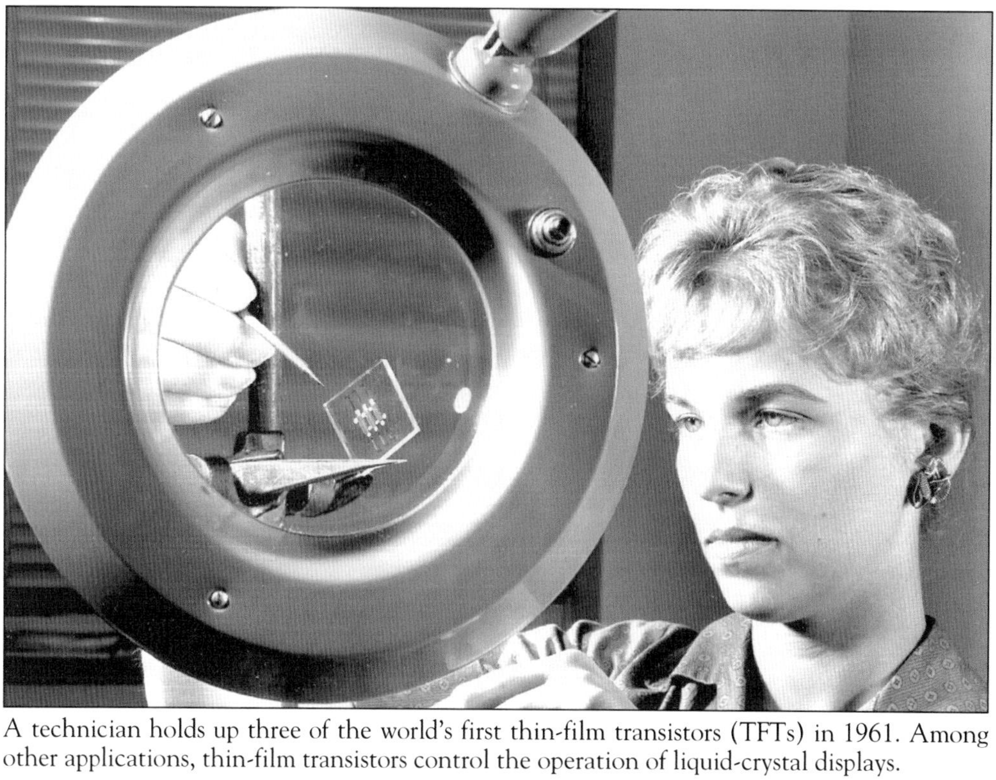

A technician holds up three of the world's first thin-film transistors (TFTs) in 1961. Among other applications, thin-film transistors control the operation of liquid-crystal displays.

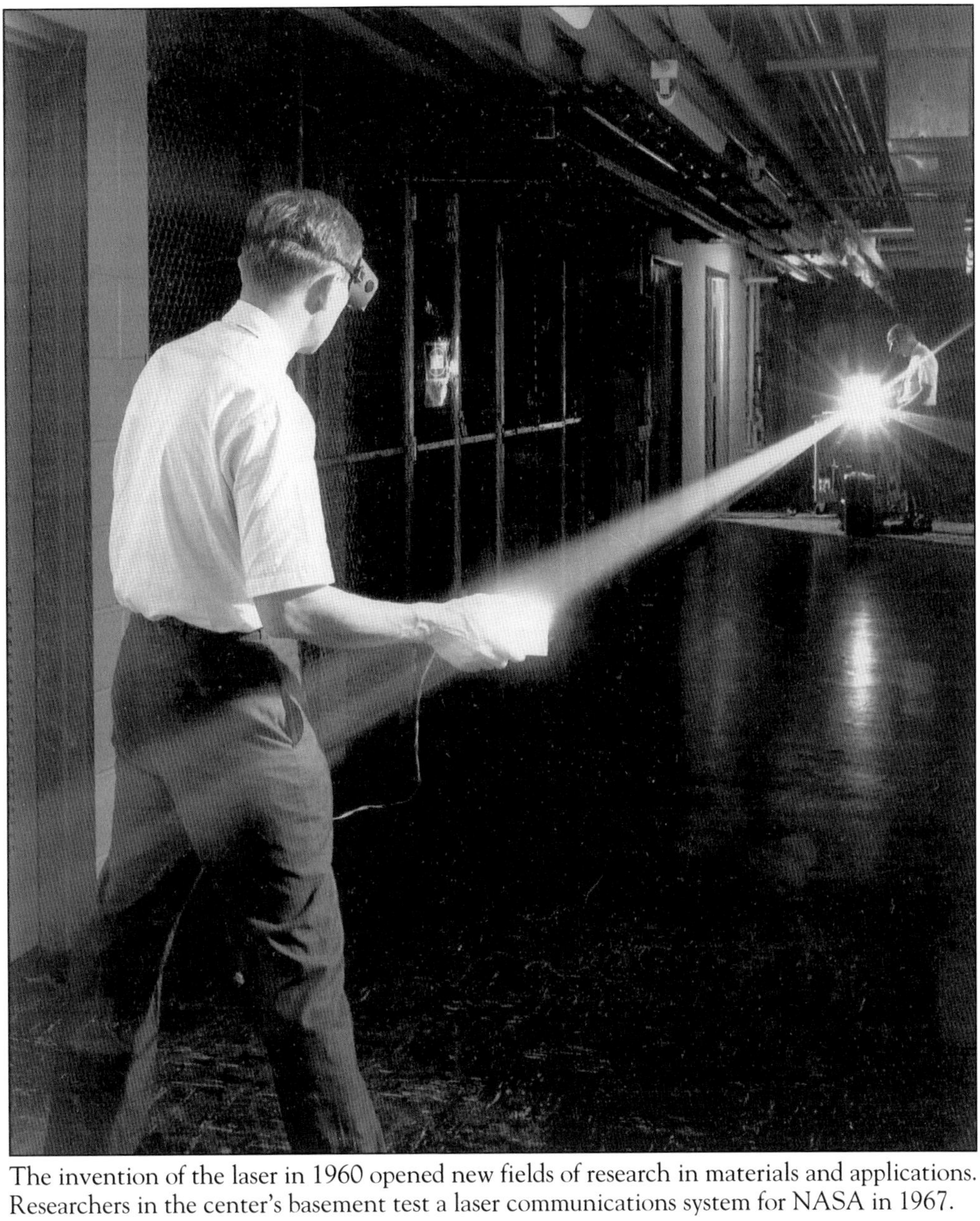

The invention of the laser in 1960 opened new fields of research in materials and applications. Researchers in the center's basement test a laser communications system for NASA in 1967.

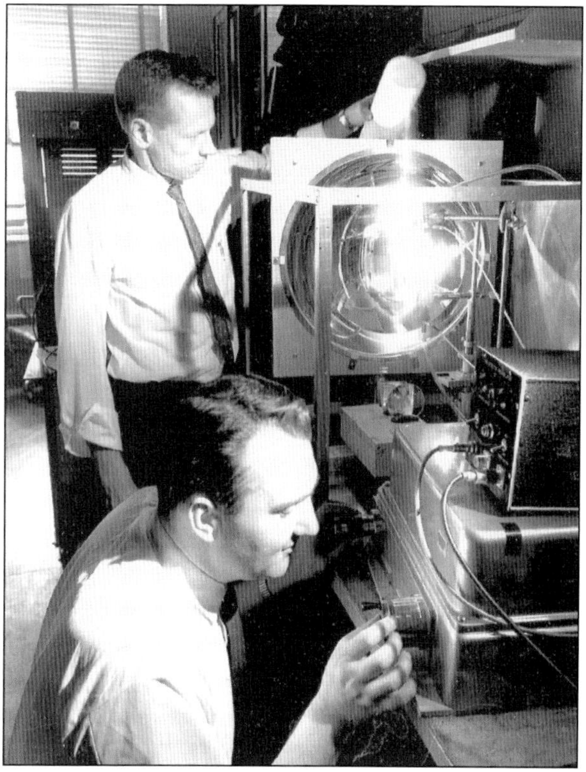

In 1962, Robert Duncan and Zoltan Kiss measure the performance of their calcium fluoride crystal laser doped with dysprosium.

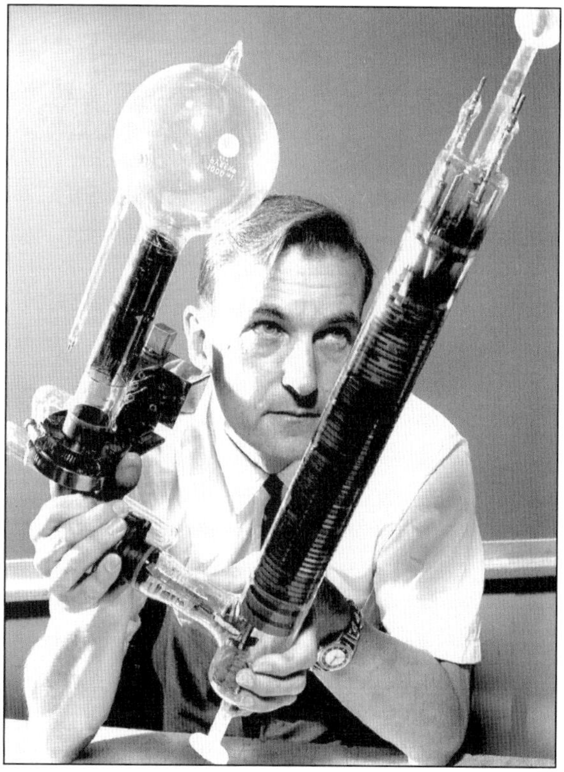

Karl Hernqvist holds up one of the first high-power lasers to operate effectively in the ultraviolet spectrum.

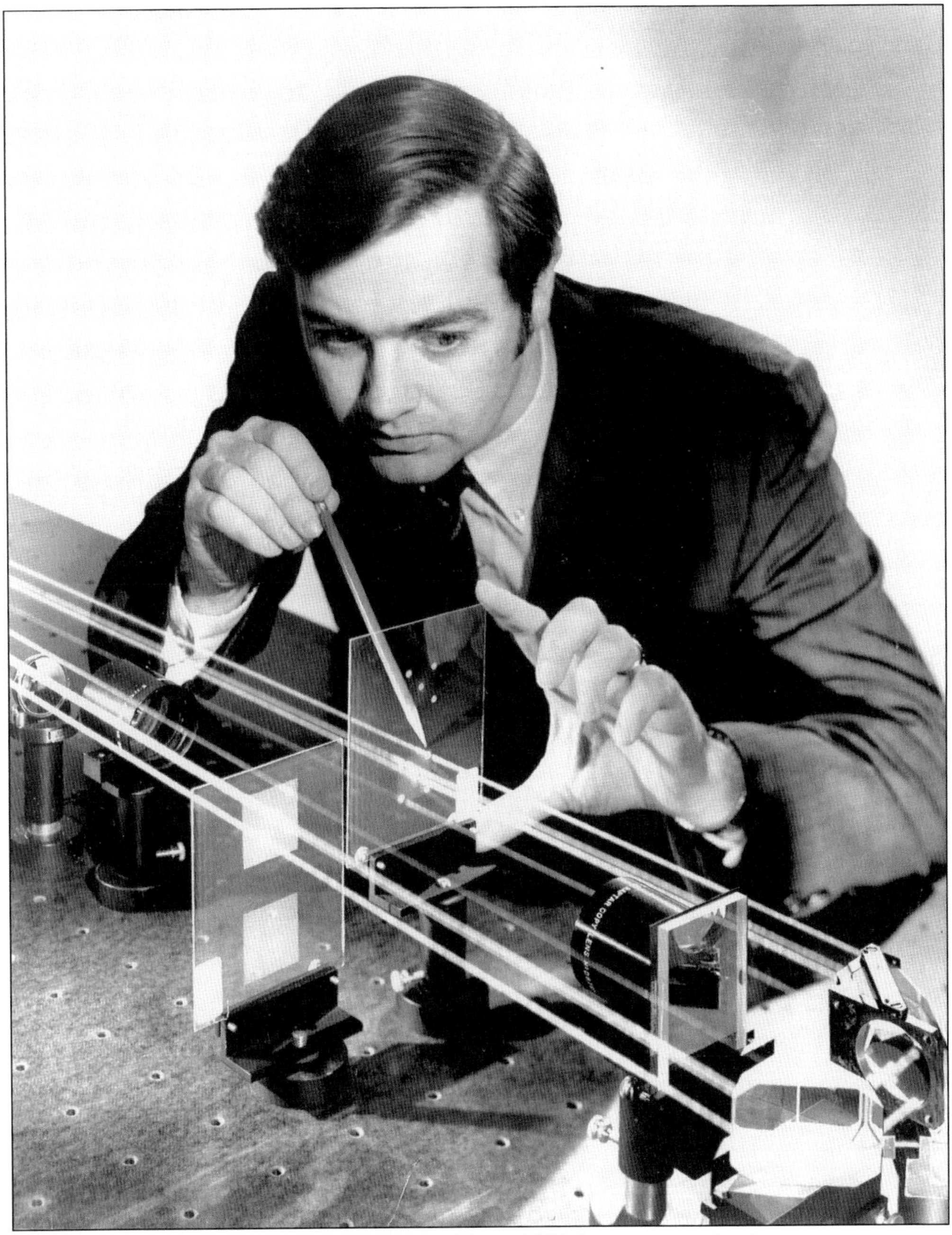

Will Stewart watches the angles on a laser table in 1971 for a communications system.

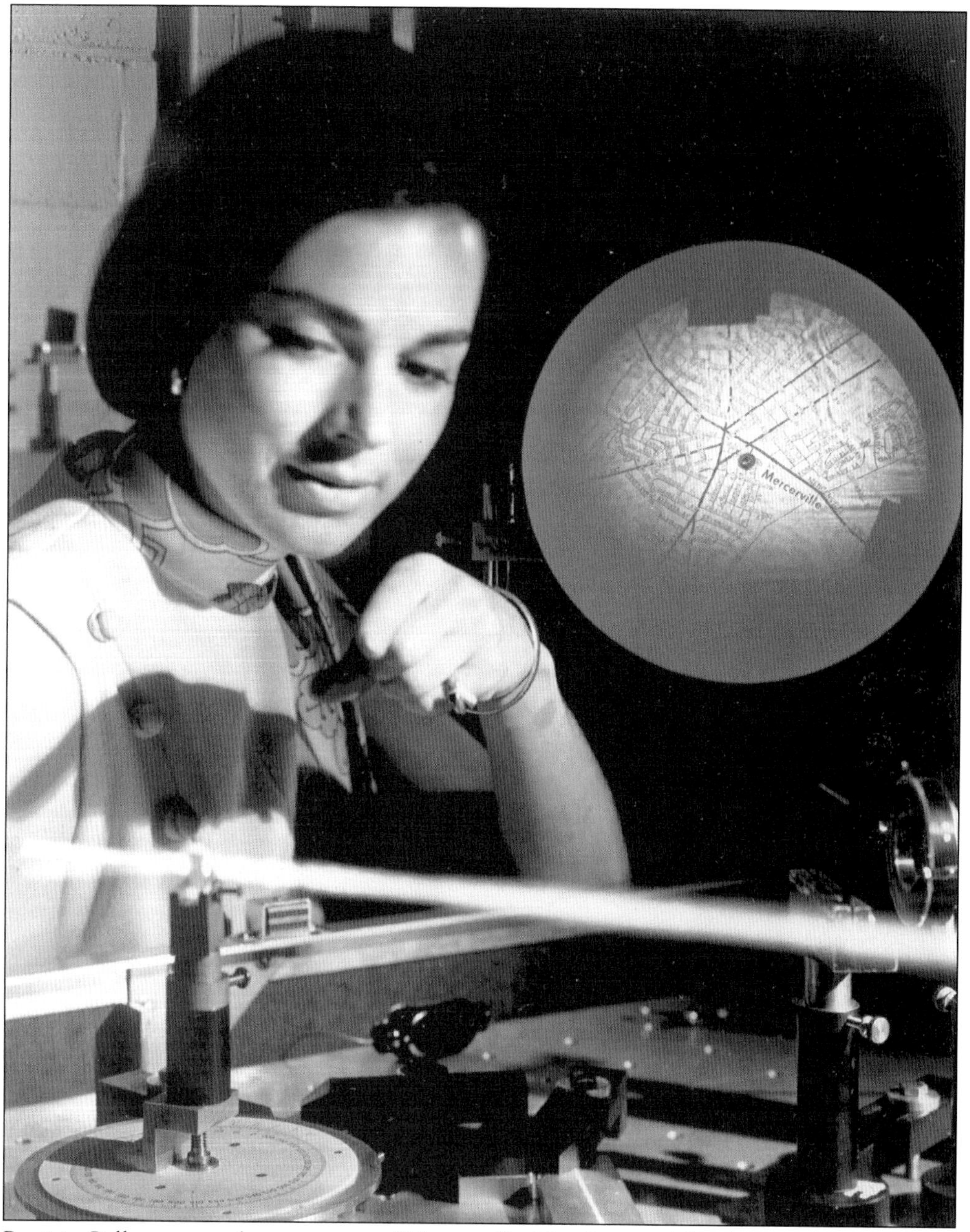

Patricia Cullen can read a street map of Mercerville, New Jersey, if she chooses, from the laser-read crystal cube in 1971.

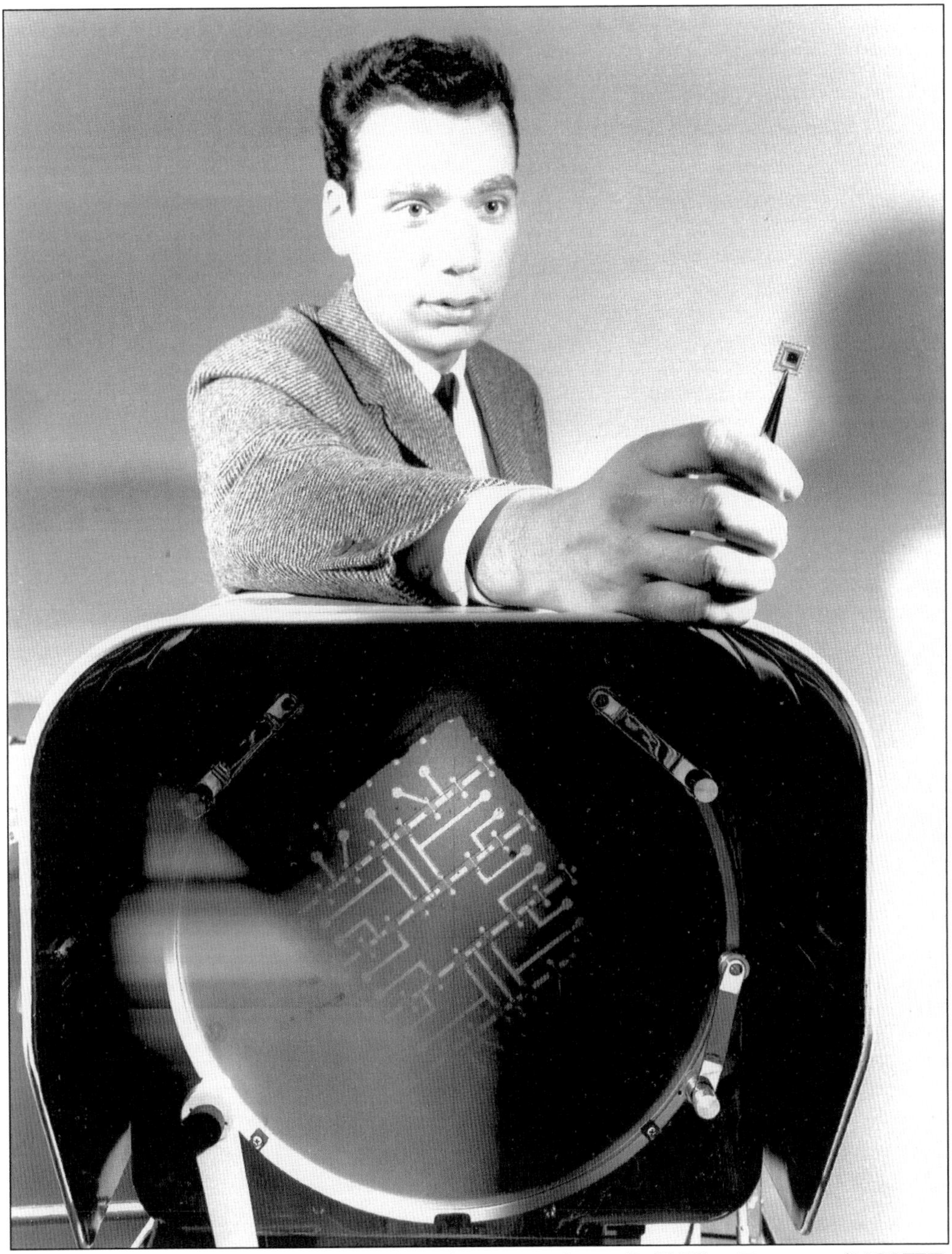

Steven Hofstein holds out the first Metal Oxide Semiconductor Field Effect Transistor (MOSFET, or MOS) circuit in 1964. Cheap and efficient, these transistors are used in virtually every electronic device and circuit and may be the most widely manufactured item in human history.

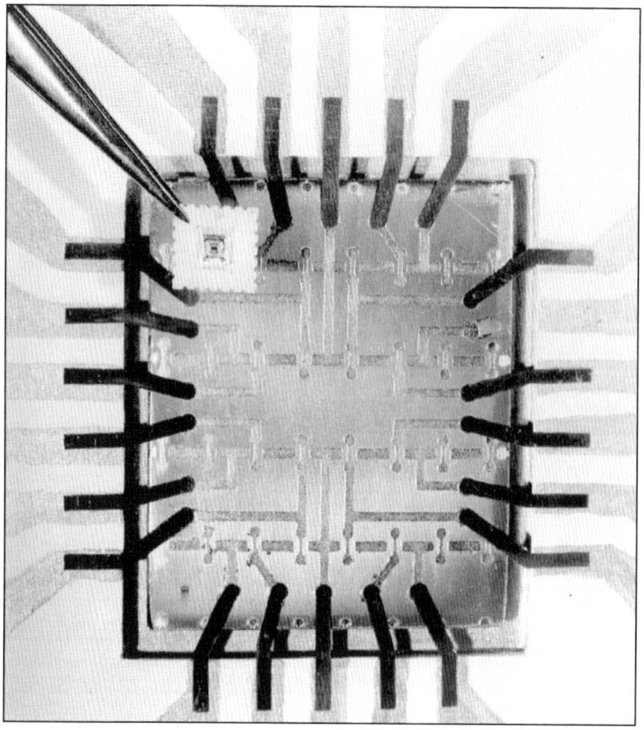

A MOS chip, with its 16 transistors on a ceramic substrate held by tweezers, is contrasted with a model.

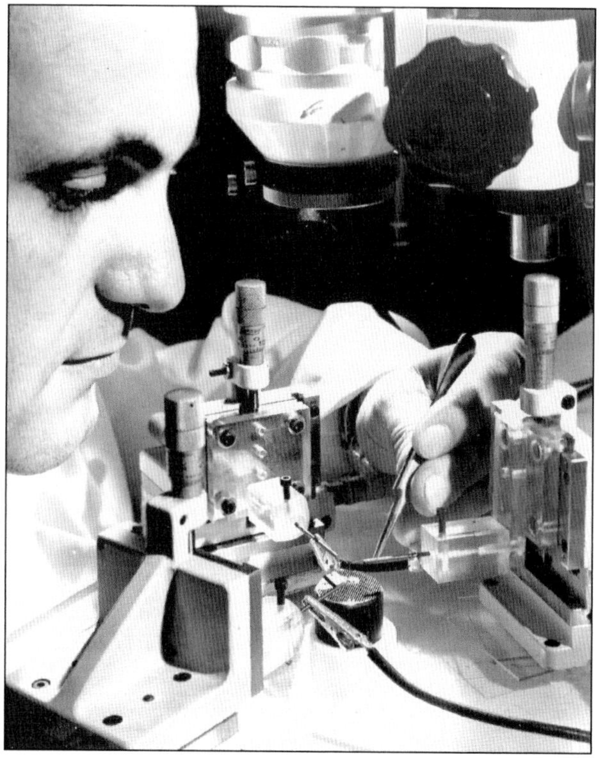

Robert Vannozzi tests a silicon wafer holding 2,200 MOS transistors in 20 complete circuits.

BOOK PAGE
15811- 78

Photograph of domain structure of p-azoxyanisole in electric field

liquid crystal between crossed polaroids. Electric field appli[ed] to upper half by means of Tic Stripe on glass plate. applied signal was 20 V peak to peak 1KC. Approximate thickness of liquid crystal was 0.1 mm. ∴ field ≈ 2×10^3

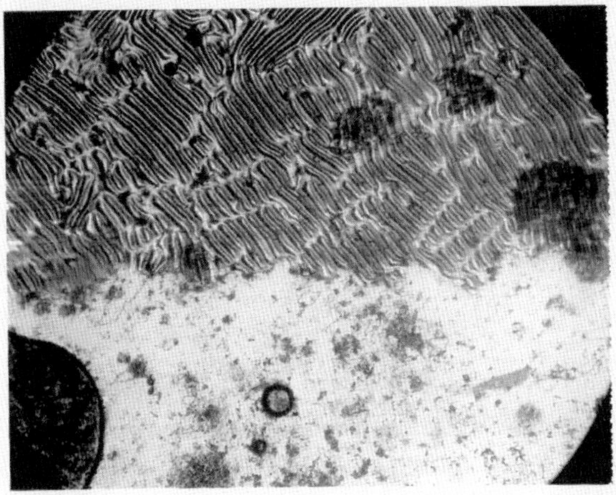

magnification 90 X

WORK OF: Richard Williams DATE: April 13, 1962

I WITNESS THIS DOCUMENT.		I WITNESS THIS DOCUMENT.	
I understand its contents: RB (Initial If Yes)	I saw the tests made: RB (Initial If Yes)	I understand its contents: R.E.S. (Initial If Yes)	I saw the tests made: R.E.S. (Initial If Yes)
NAME: R. Bruder	DATE: 4/30/62	NAME: R E Shrader	DATE: 4-30-62

Looking for a way to make a television to hang on the wall, chemist Richard Williams put some liquid crystals between two panes of glass coated with a transparent conductor in 1962. This notebook page and attached photograph represent the ancestor of all liquid-crystal displays on earth and led to Williams's receipt of the first patent for liquid-crystal displays. The liquid crystals aligned under the electric field are called Williams Domains in his honor.

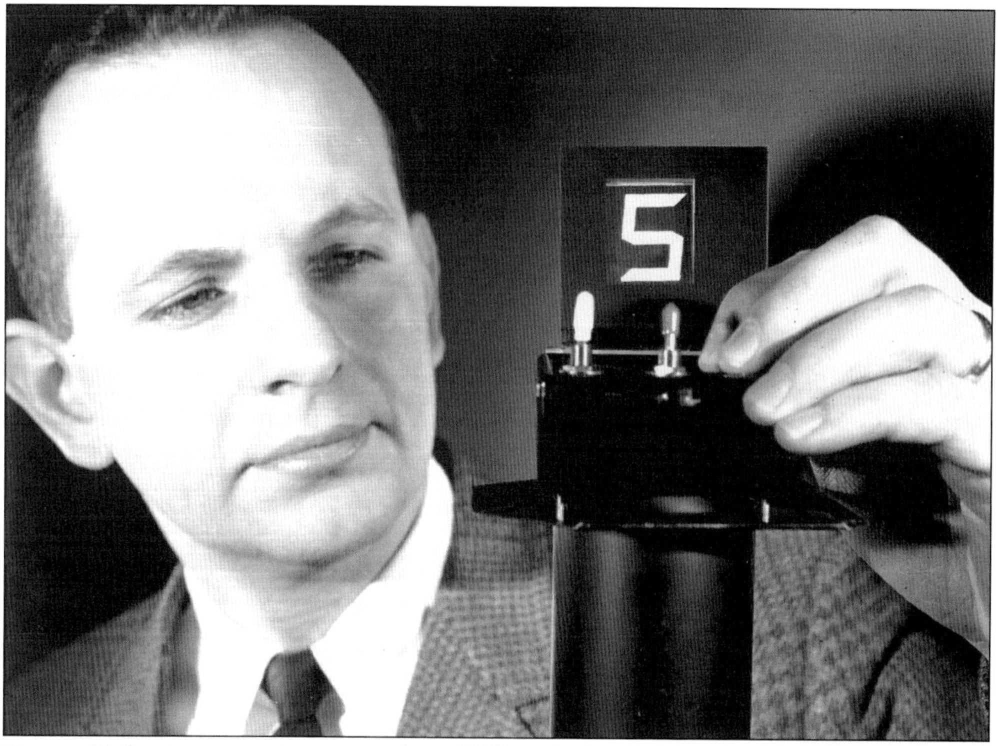

George Heilmeier spent six years making Williams's demonstration of principle a practical device, culminating in this digital counter at RCA's press conference in May 1968.

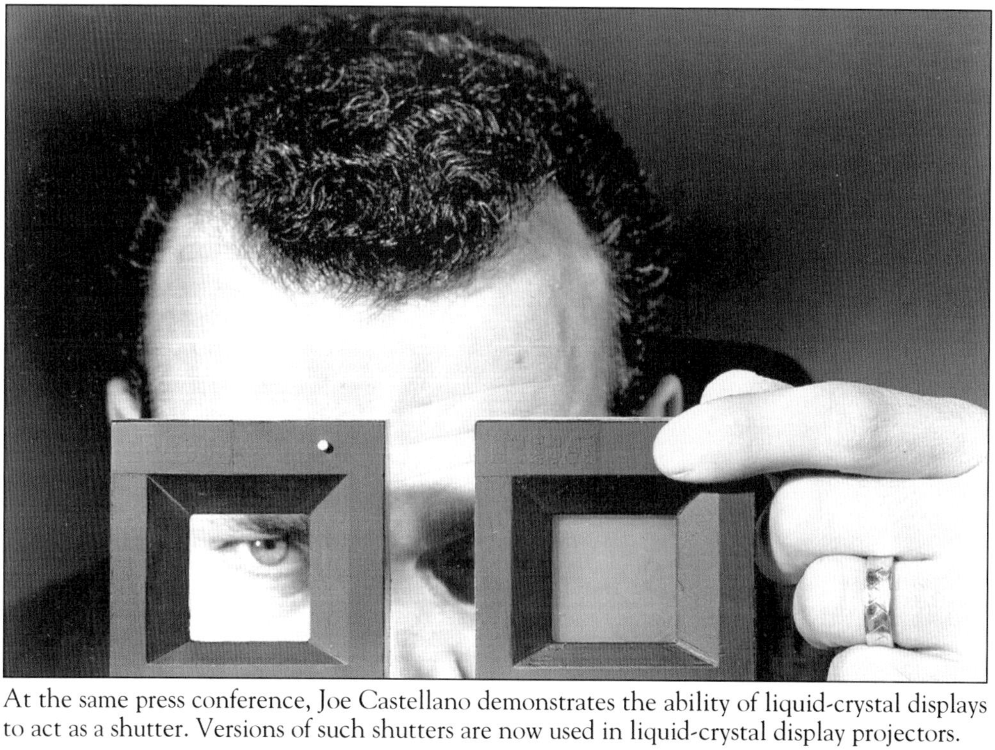

At the same press conference, Joe Castellano demonstrates the ability of liquid-crystal displays to act as a shutter. Versions of such shutters are now used in liquid-crystal display projectors.

Four

INVENTING TO THE END OF AN ERA 1971–1986

David Sarnoff died on December 12, 1971, at the age of 80. His passing marked an unofficial turn toward the center's increased cooperation with RCA's other divisions amid rising Asian competition in consumer electronics and investment funds' demands for short-term corporate profits. Those pressures, compounded by four changes in RCA's leadership over six years, frustrated innovation of liquid-crystal displays, home video cameras, solar cells, and the personal computer. Most painful of all was the commercial failure of the VideoDisc system, which provided two hours of prerecorded video stamped on a 12-inch record.

Nonetheless, the center's staff made significant improvements in color television receivers, semiconductor lasers, video cameras, and microprocessors. Other developments presaged future areas of innovation. The future looked bright in 1986, when General Electric announced that it was purchasing RCA in the largest nonoil corporate acquisition in U.S. history.

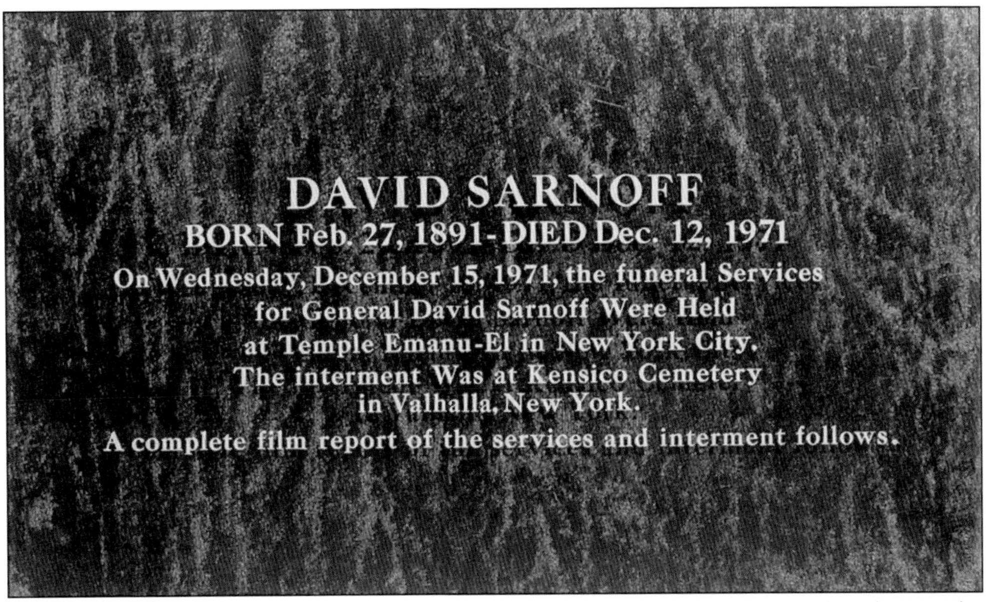

NBC broadcast this announcement before screening an obituary film of its cofounder. At the center, a memorial ceremony was held in the auditorium.

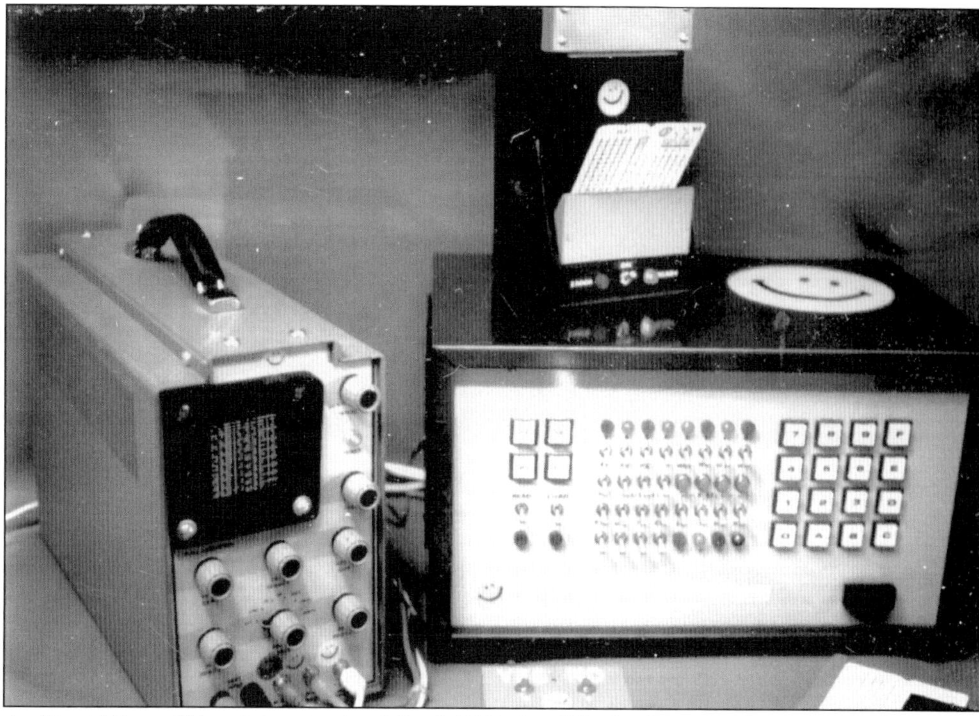

As David Sarnoff lay dying in the fall of 1971, Joe Weisbecker was designing an affordable personal computer from his home in Haddonfield, New Jersey, where he did much of his work. The 1972 Model 00 was the result: a $975 card- or cassette-fed programmable computer suitable for playing simple video games on an oscilloscope or a television set.

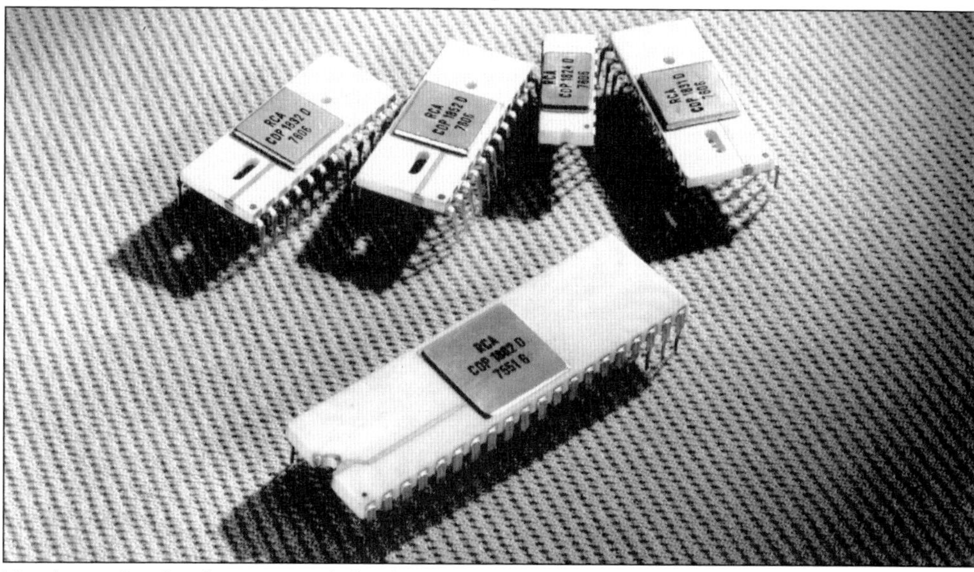

Not long after Intel announced the first microprocessor, or computer on a chip, Weisbecker sketched out the logic for the first CMOS (Complementary Metal Oxide Semiconductor Field Effect Transistor) microprocessor. The RCA 1802, announced in 1976, was applied to everything from Weisbecker's Studio II game computer, to three years of certain Chrysler automobiles, to the Voyager spacecraft, which continue to process data 26 years after launching.

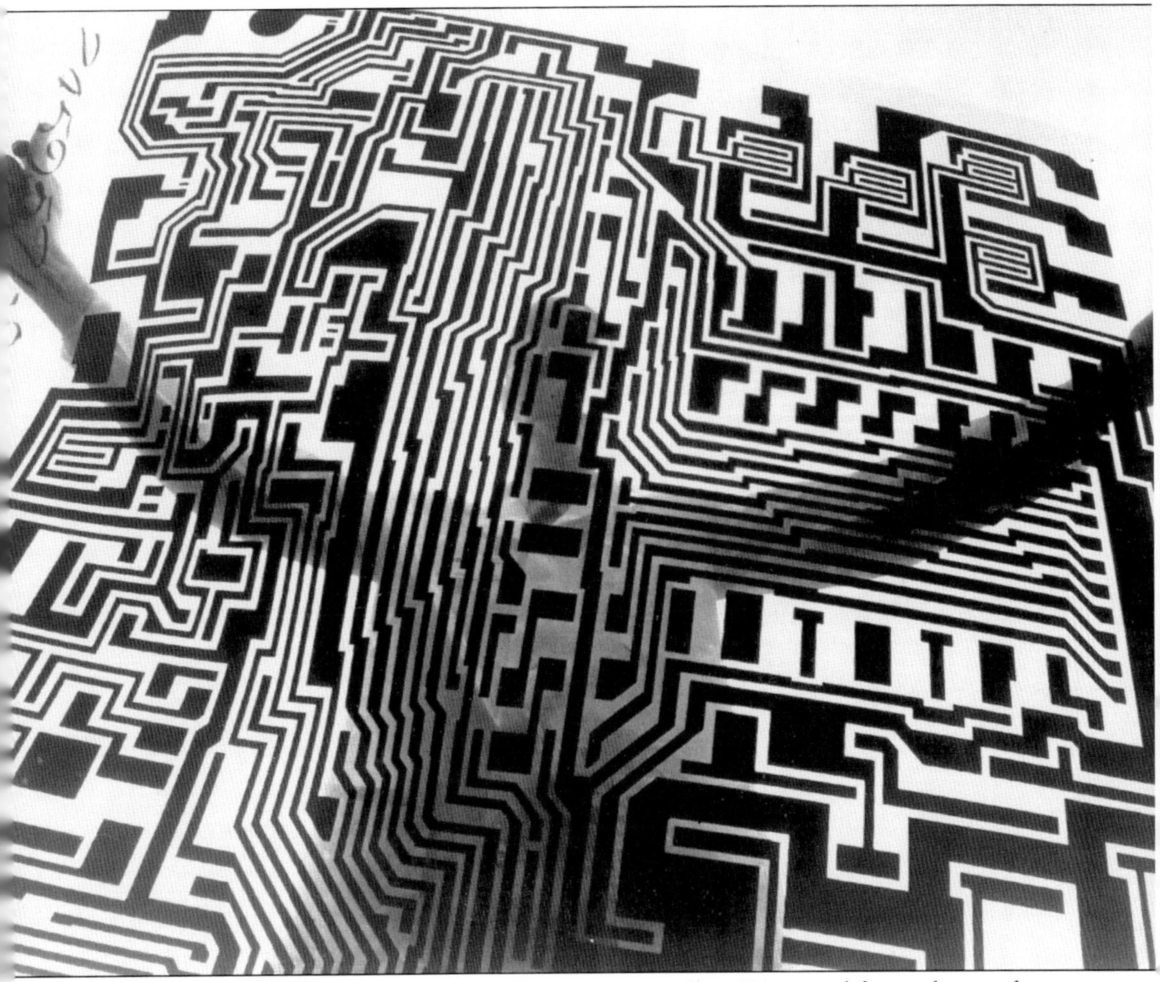

A member of the center's integrated circuit department unrolls a 500-times life-size design of an integrated circuit in 1975.

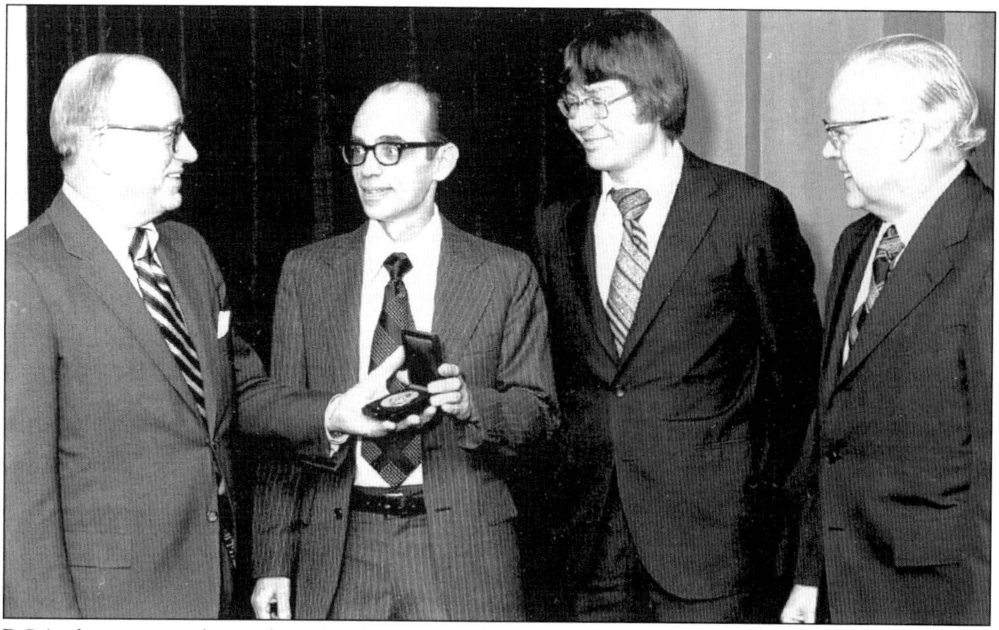

RCA chairman and president Anthony Conrad (far left) and RCA Laboratories vice president James Hillier (far right) congratulate Joe Weisbecker (second from left) and his supportive supervisor, Robert Winder. They received the 1976 David Sarnoff Award for Outstanding Technical Achievement "for excellence of team effort leading to the development and marketing of an advanced microprocessor," that is, the CMOS microprocessor.

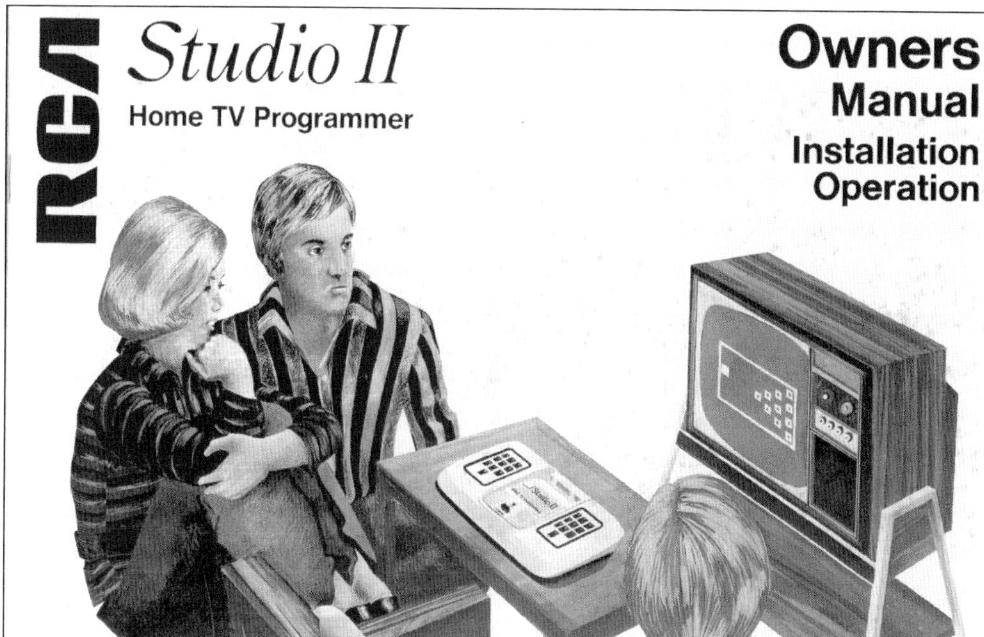

That the artwork on this initial manual for the 1802-driven Studio II game computer was ever released suggests that some in RCA's management were less than enthusiastic about Weisbecker's home entertainment alternative to prerecorded video.

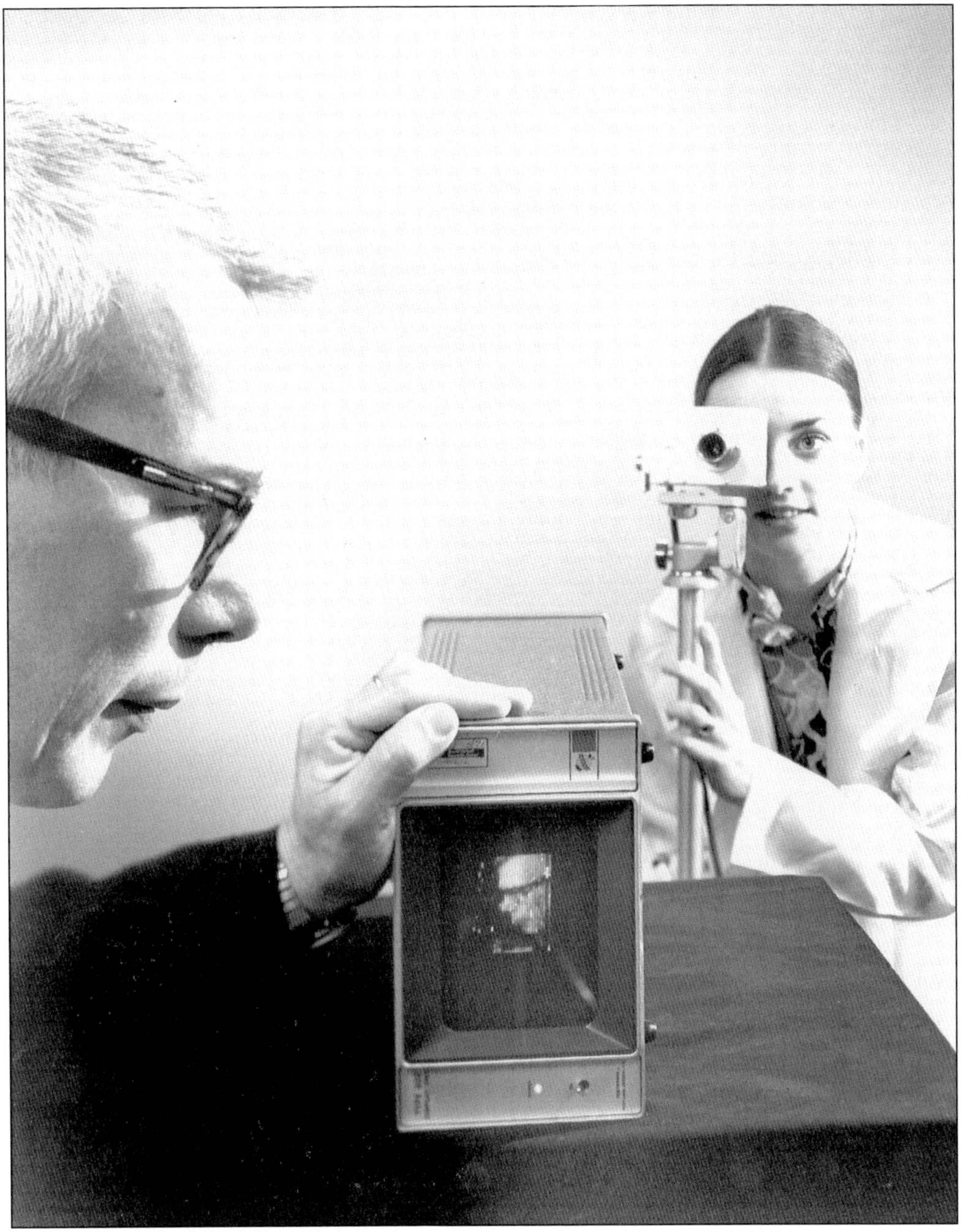

Lynn Roberts shoots Win Pike with the new charge-coupled device (CCD) camera in March 1972. Bell Labs researchers invented the CCD chip, but the center's staff and RCA's Broadcast Division were the first to bring it to market, 12 years later, in an Emmy-winning camera.

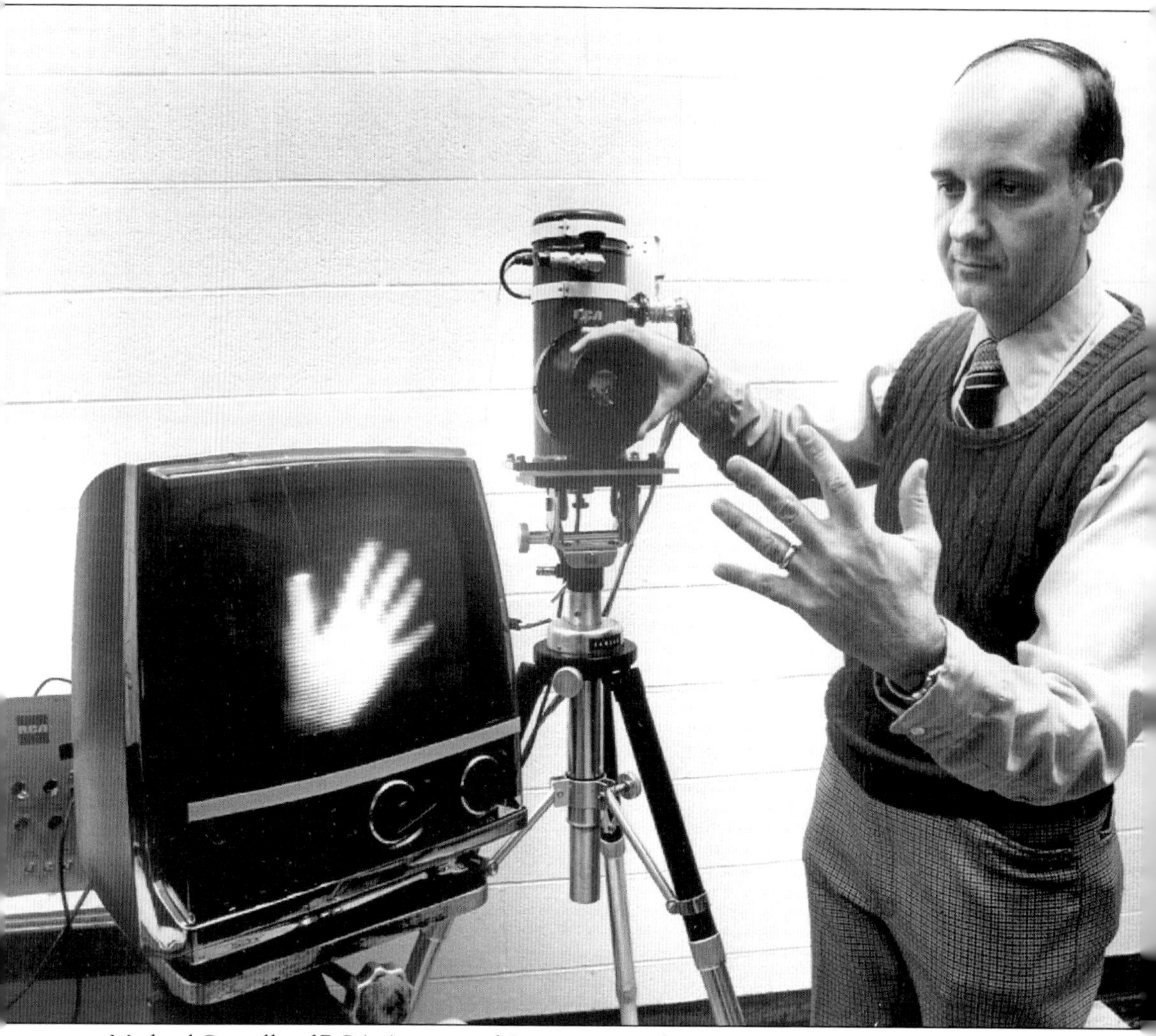

Michael Cantella of RCA Automated Systems in Burlington, Massachusetts, shows the heat in his hand and its blockage by his sleeve on an infrared CCD camera. The image chip developed at the center marked an extension of innovations in infrared tube video started during World War II. Infrared imagers play critical roles in search-and-rescue and military operations.

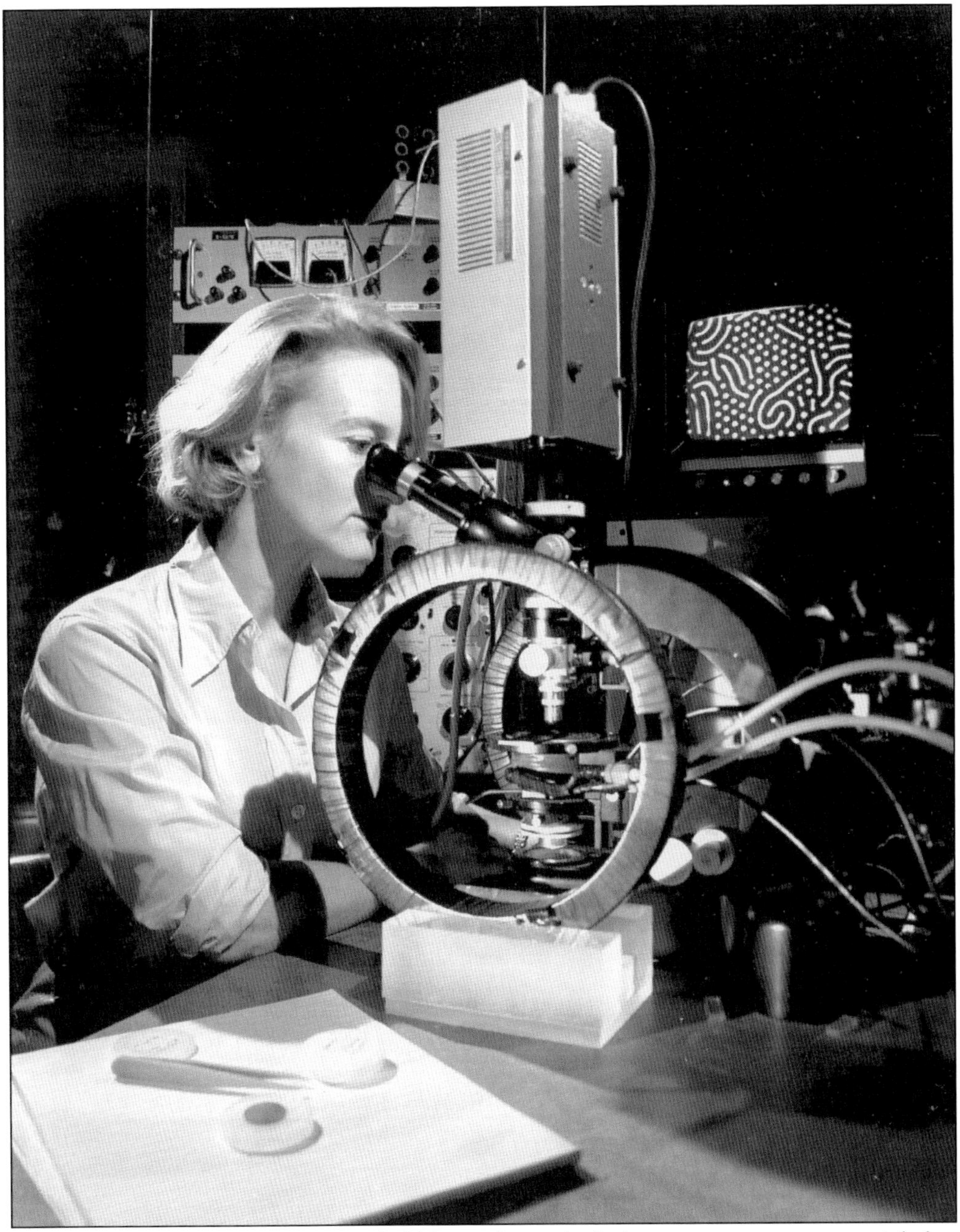

Aline Akselrad examines magnetic bubbles in bismuth thulium garnet, a new material that increases the speed of data readout in computer bubble memories. The bubbles appear as dots on the video screen.

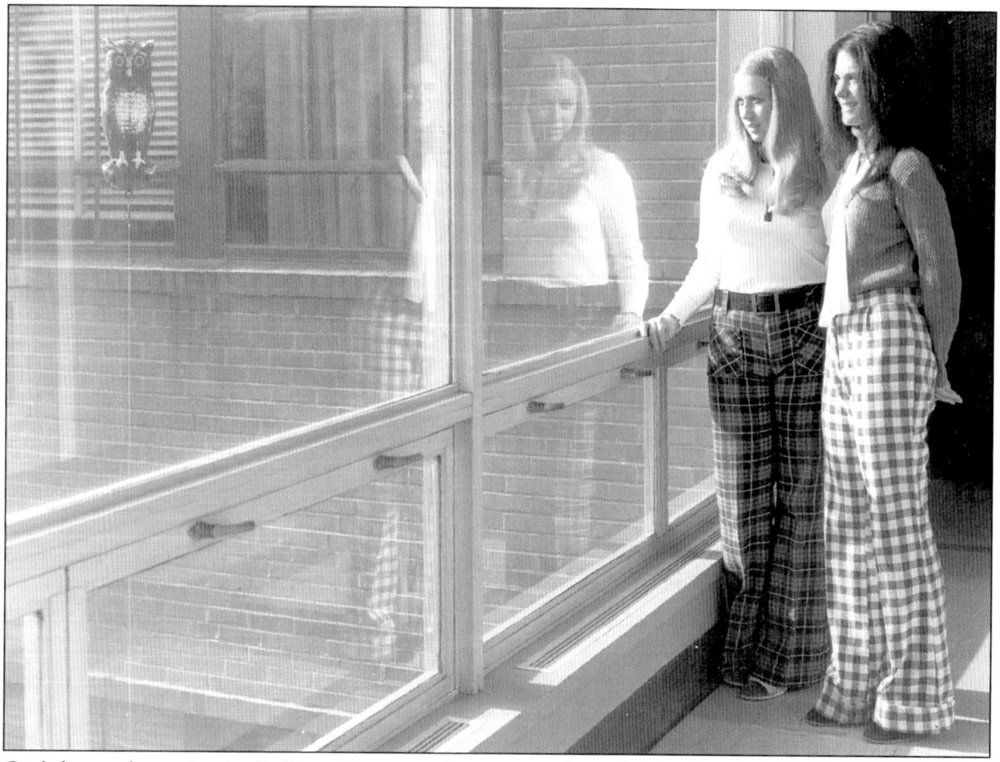

Owl decoys hung in the hallway between the west and southwest wings during the early 1970s were intended to persuade other birds not to fly into the windows.

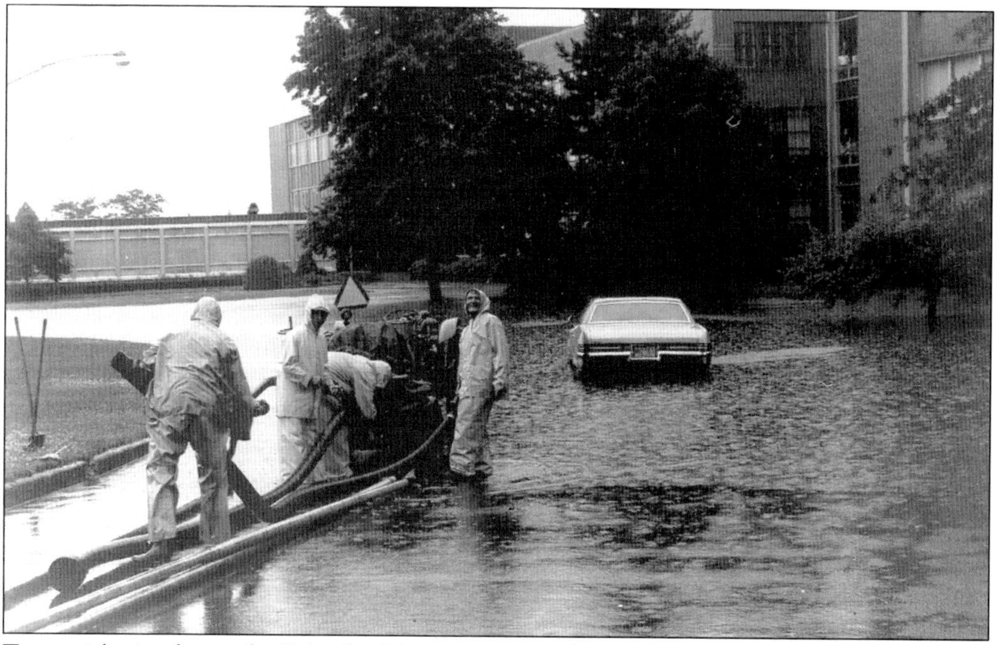

Torrential rains during the "Monday Monsoon" in April 1975 overwhelmed the drainage system in front of the southwest wing. The facilities crew prepares to drain 18 inches of water while concerned staff members look on from the hallways.

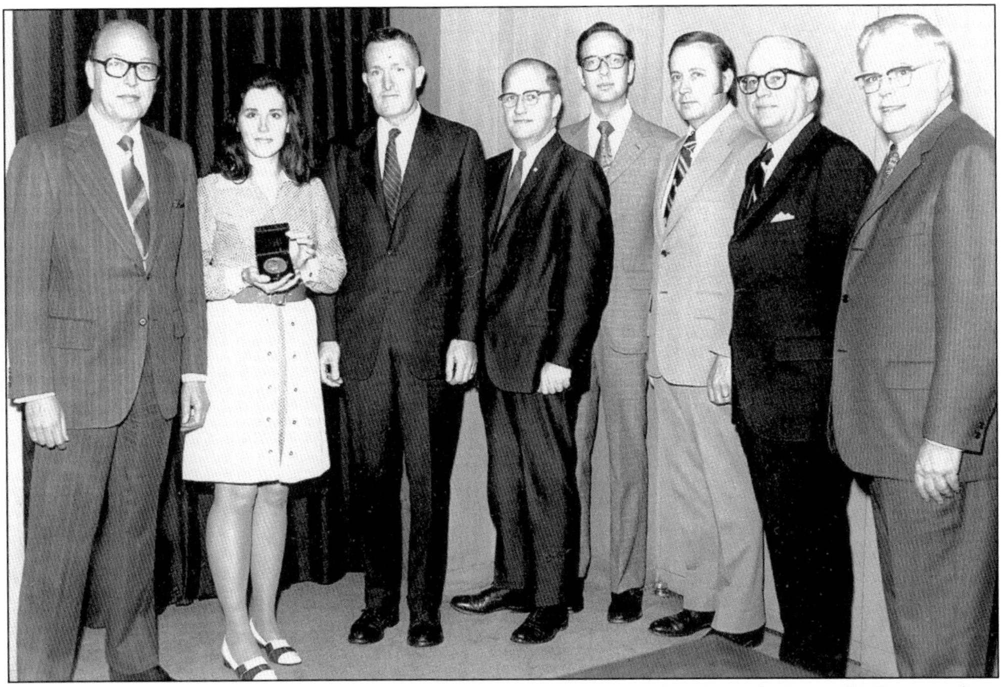

In 1972, RCA chief executive Robert Sarnoff presents the David Sarnoff Outstanding Achievement Award to William Hannan's team for their innovation of a holographic videotape system. To the right of the unidentified woman are, from left to right, William J. Hannan, Robert E. Flory, Michael Lurie, Dainis Karlsons, Robert J. Ryan, and James Hillier (vice president for RCA Laboratories).

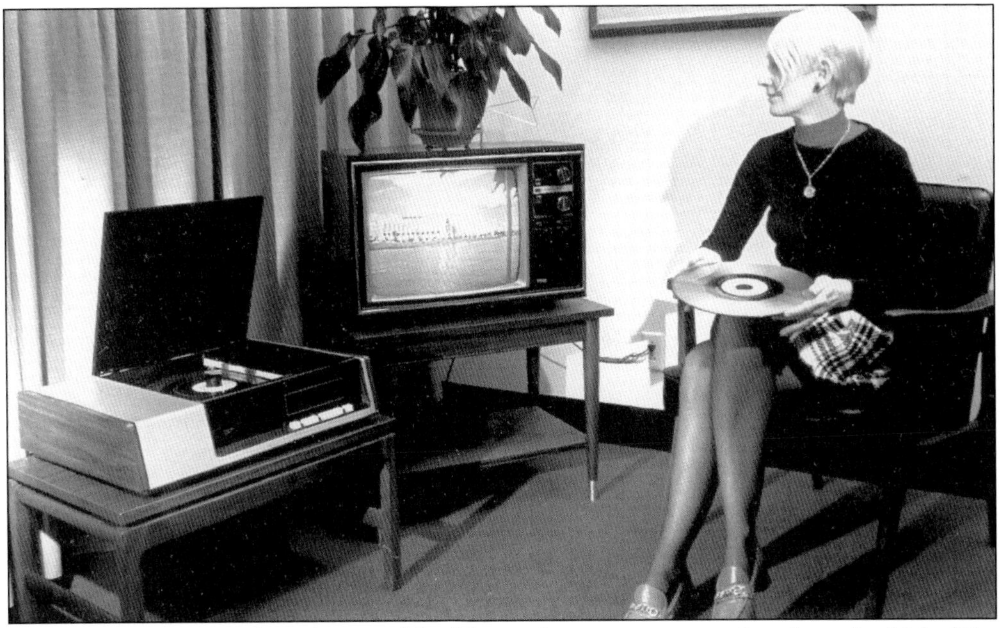

The SelectaVision trademark started by Hannan's group was transferred to the VideoDisc project in 1973. The initial hope was that consumers would be able to treat the VideoDisc as a viewable phonograph record.

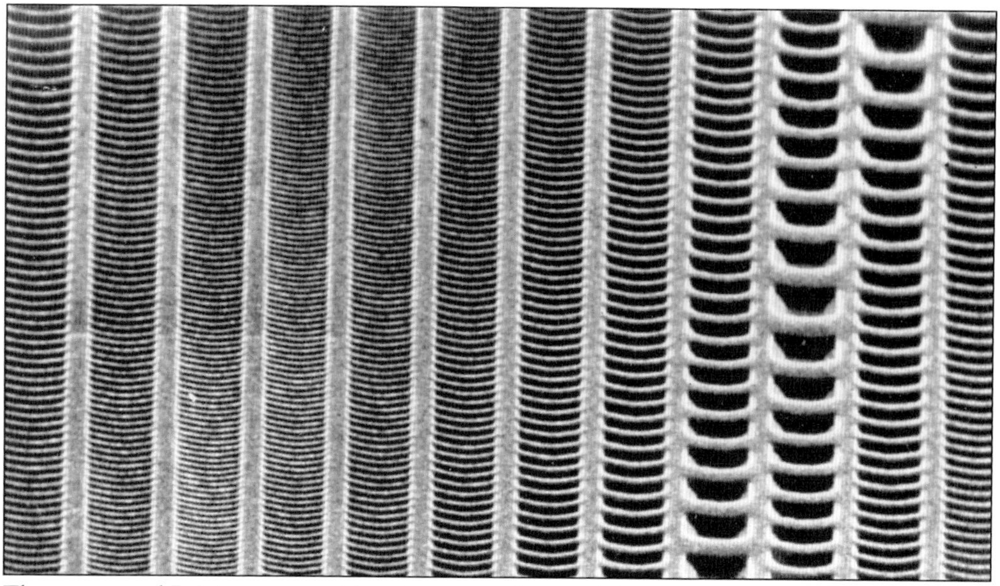

The center and RCA committed hundreds of millions of dollars to recording and reproducing video in grooves packed nearly 10,000 to the inch. This electron micrograph shows the recording by electron beam in some of those grooves.

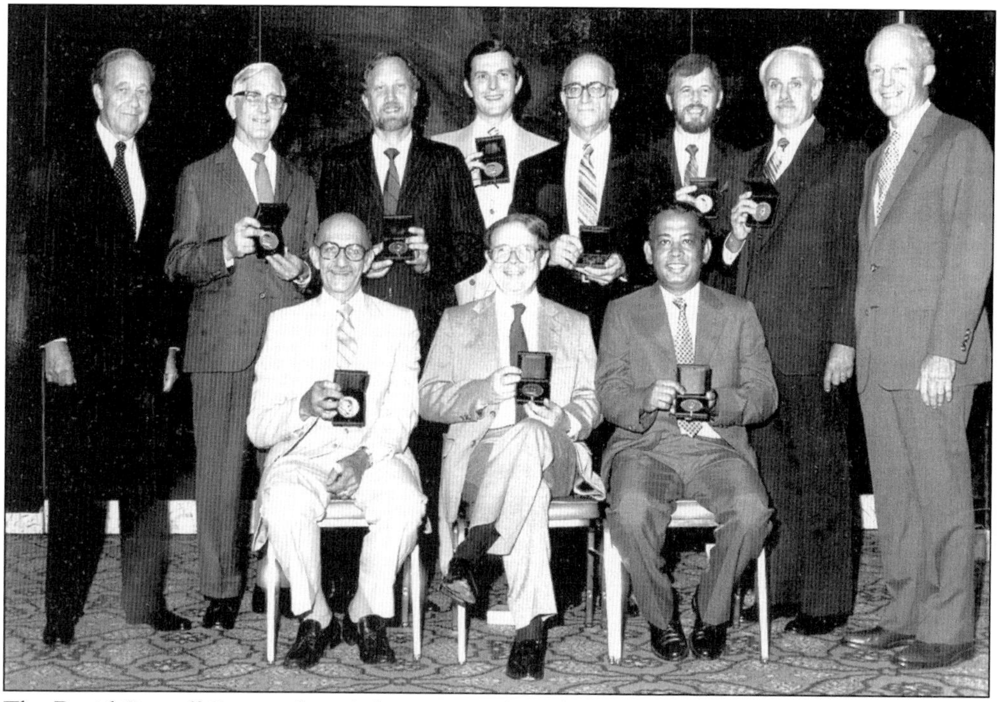

The David Sarnoff Outstanding Achievement Awards in 1982 went to the innovators of the most expensive consumer entertainment innovation to that time. From left to right are the following: (front row) Leonard Fox, Marvin Leedom, and Pabitra Datta; (back row) RCA chairman Thornton Bradshaw, Eugene Keizer, Jon Clemens, Todd Christopher, Jerome Halter, Michael Miller, Frederick Stave, and executive vice president of research and engineering William Hittinger.

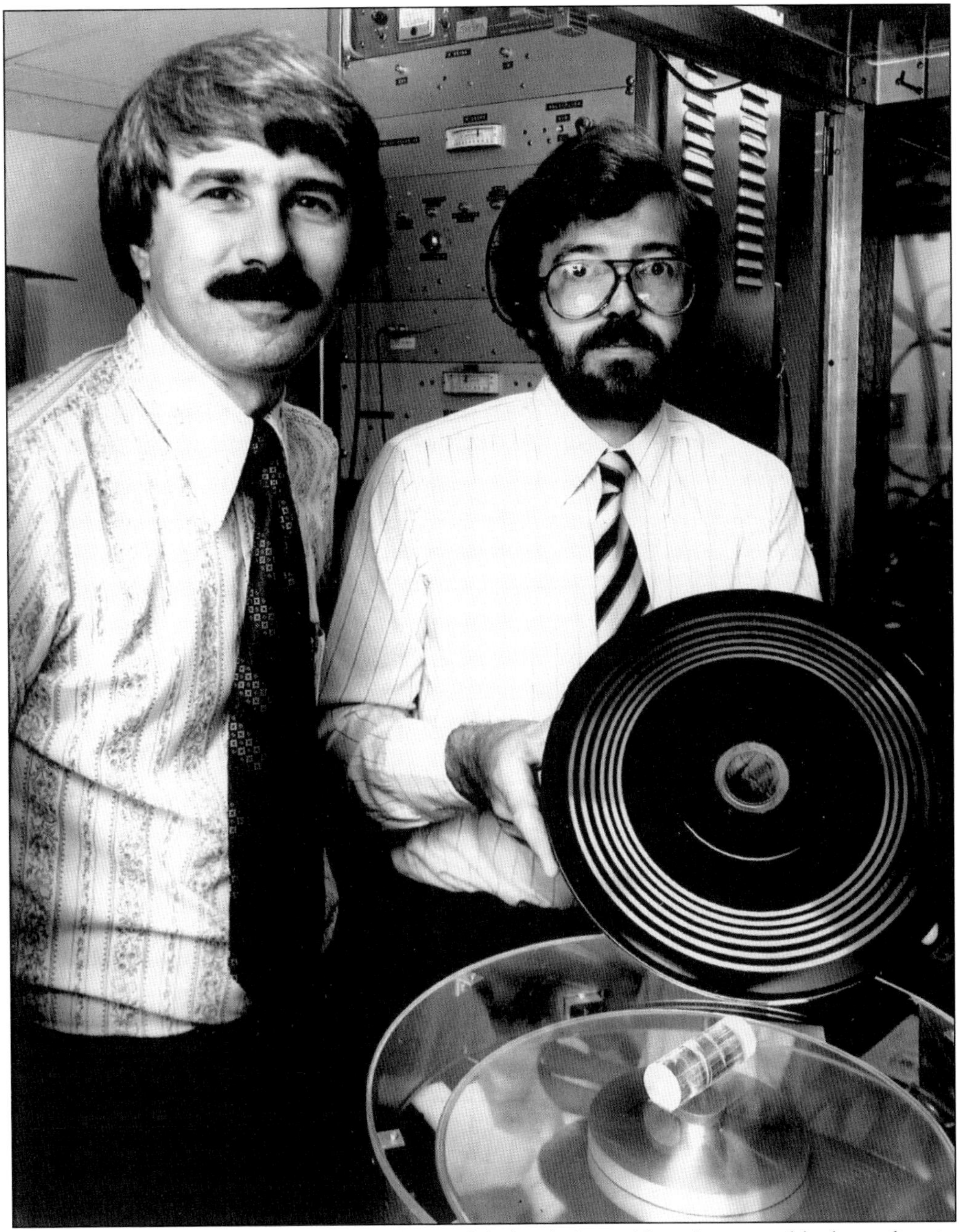

Robert Bartolini (left) and Alan Bell stand behind their 12-inch rewritable laser disc in 1980. A semiconductor laser beam invented by Dan Botez stores and erases over 12 gigabytes of information on the disc, while Fred Spong's inventions link the laser efficiently to the information on the disc.

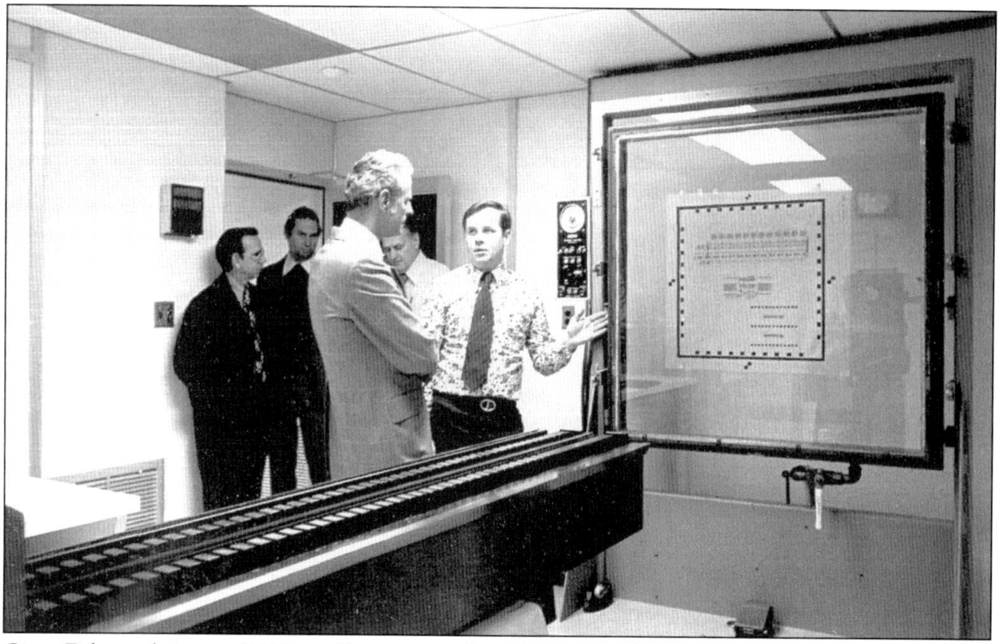

Greg Zak explains to RCA staff vice president Peter Friederich the camera used to reduce integrated-circuit designs in the center's new Class-100 white room in 1978. With fewer than 100 particles half a micron or less per cubic foot of air, the room reduces the risk of damage to circuit board artwork and integrated-circuit masks for chip fabrication.

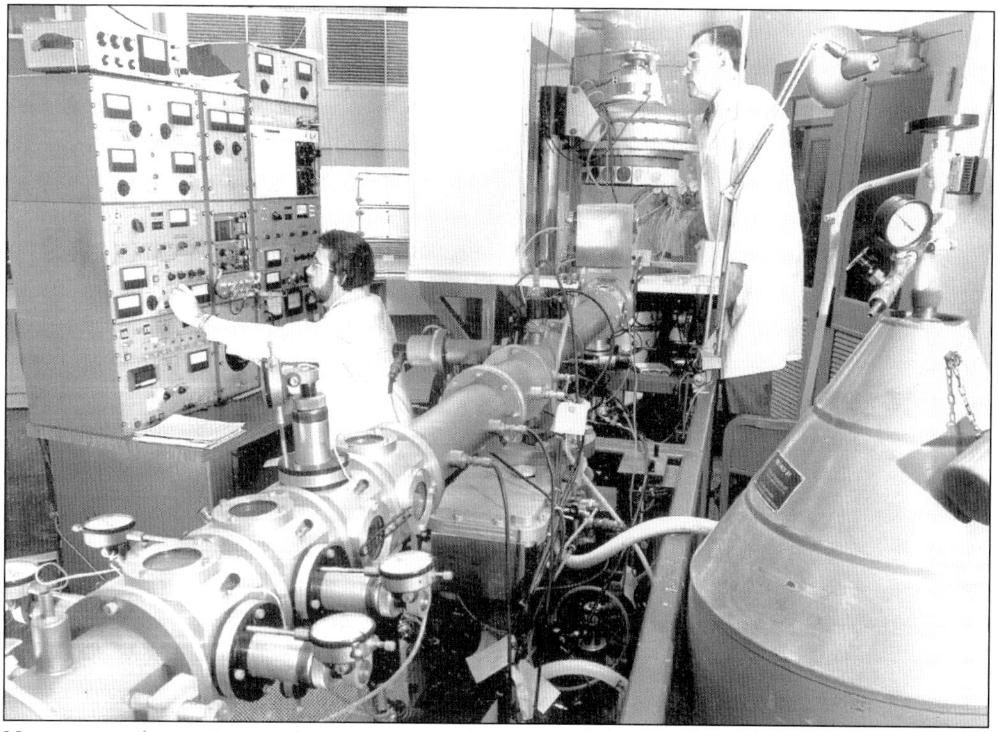

Here we see the equipment for precision implantation of the ions that "dope" silicon wafers for more efficient integrated circuits.

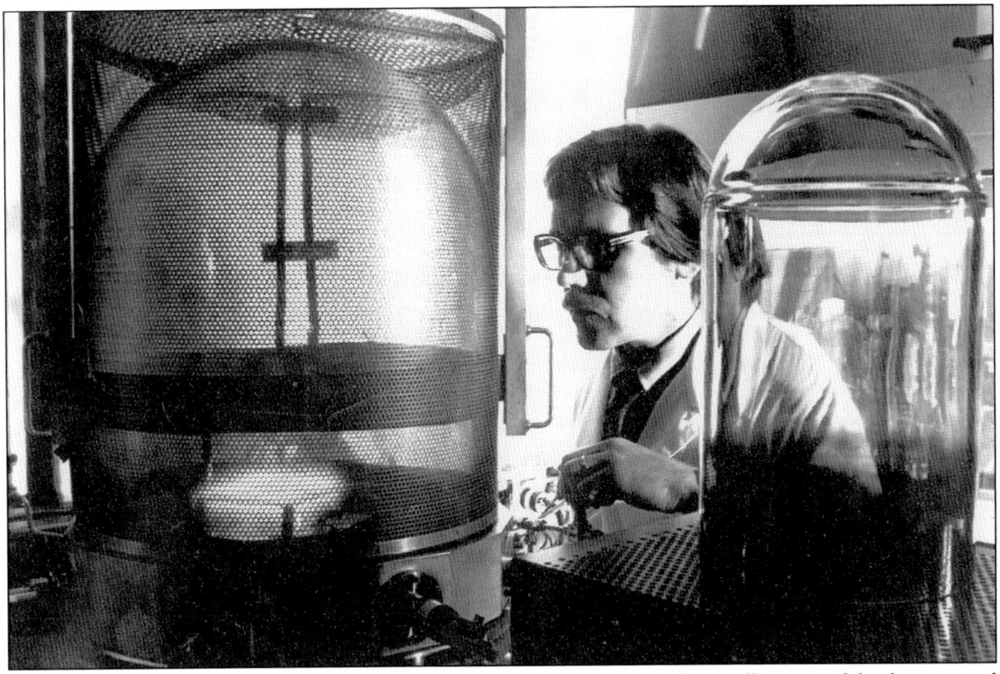

David Carlson sets up glow-discharge equipment to make solar cells out of hydrogenated amorphous silicon in 1978, for which he received the basic patent.

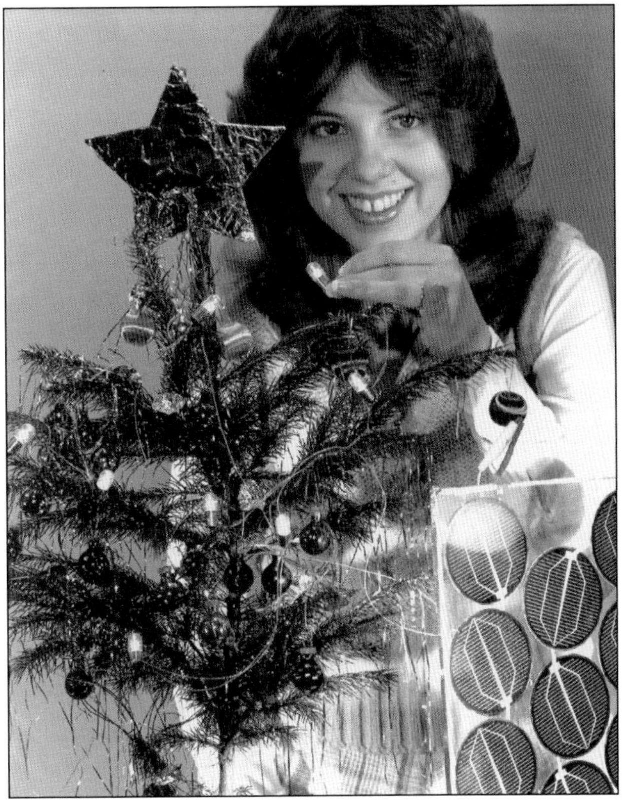

Cheryl Endler lights a Christmas tree in 1980 at the center with the most efficient solar cells ever developed. The cells converted 10 percent of sunlight into electricity and made the technology practical for more than satellite applications.

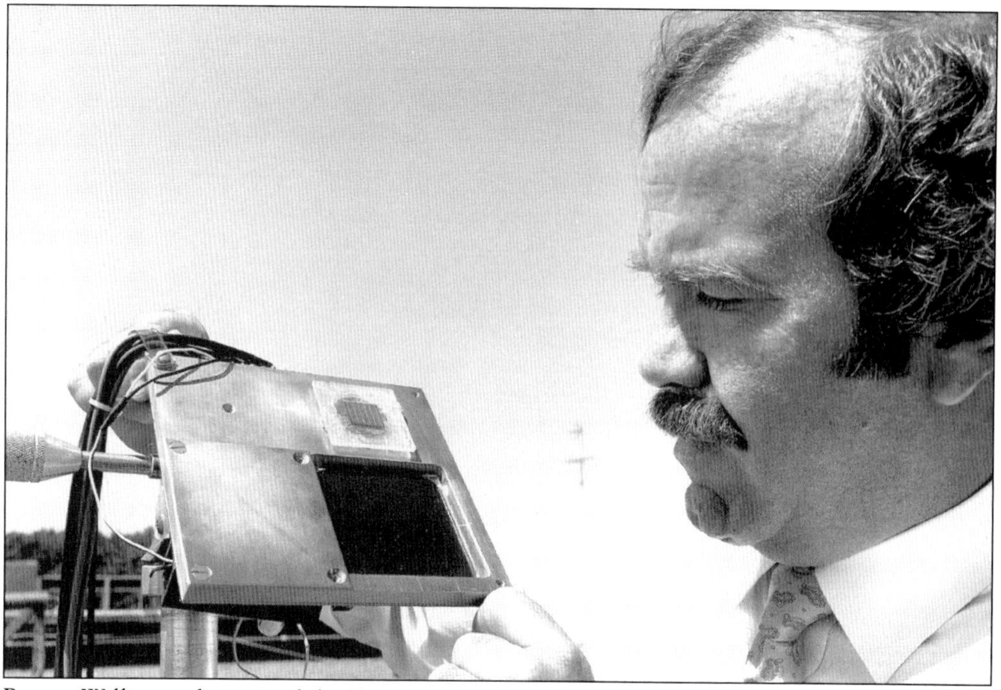

Brown Williams, director of the Energy Systems Laboratory, examines a set of solar cells on the roof of the center that are covered with lenses to concentrate more sunlight. Richard Marx and Louis Napoli designed the system to automatically position itself for maximum energy conversion.

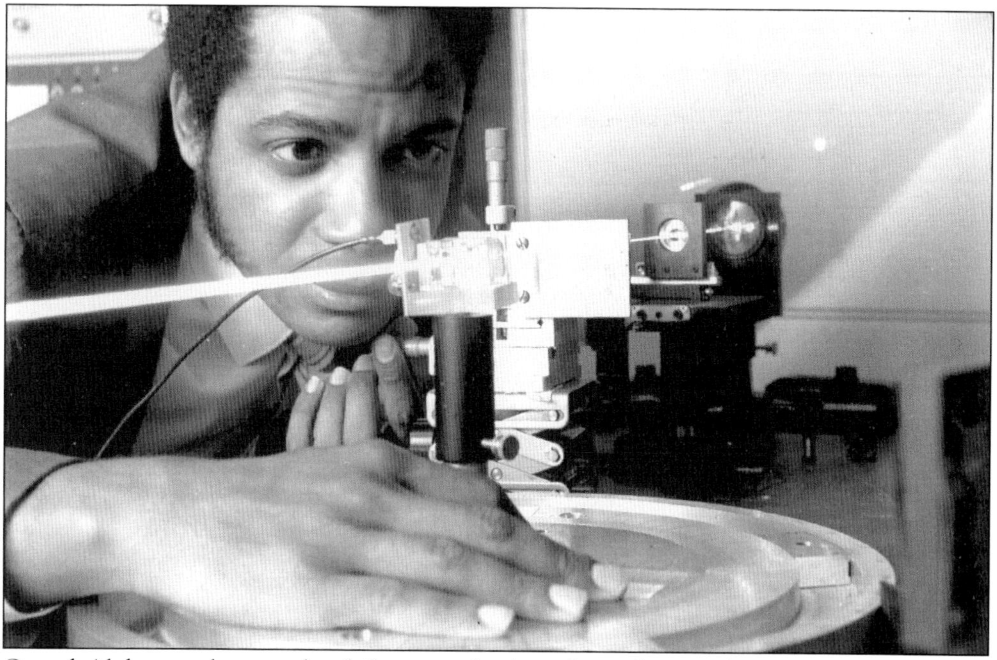

Gerard Alphonse observes the deflection of a laser beam by sound in November 1972. A transducer converts electrical energy to acoustic waves to deflect a laser beam as a means of projecting high-definition video information.

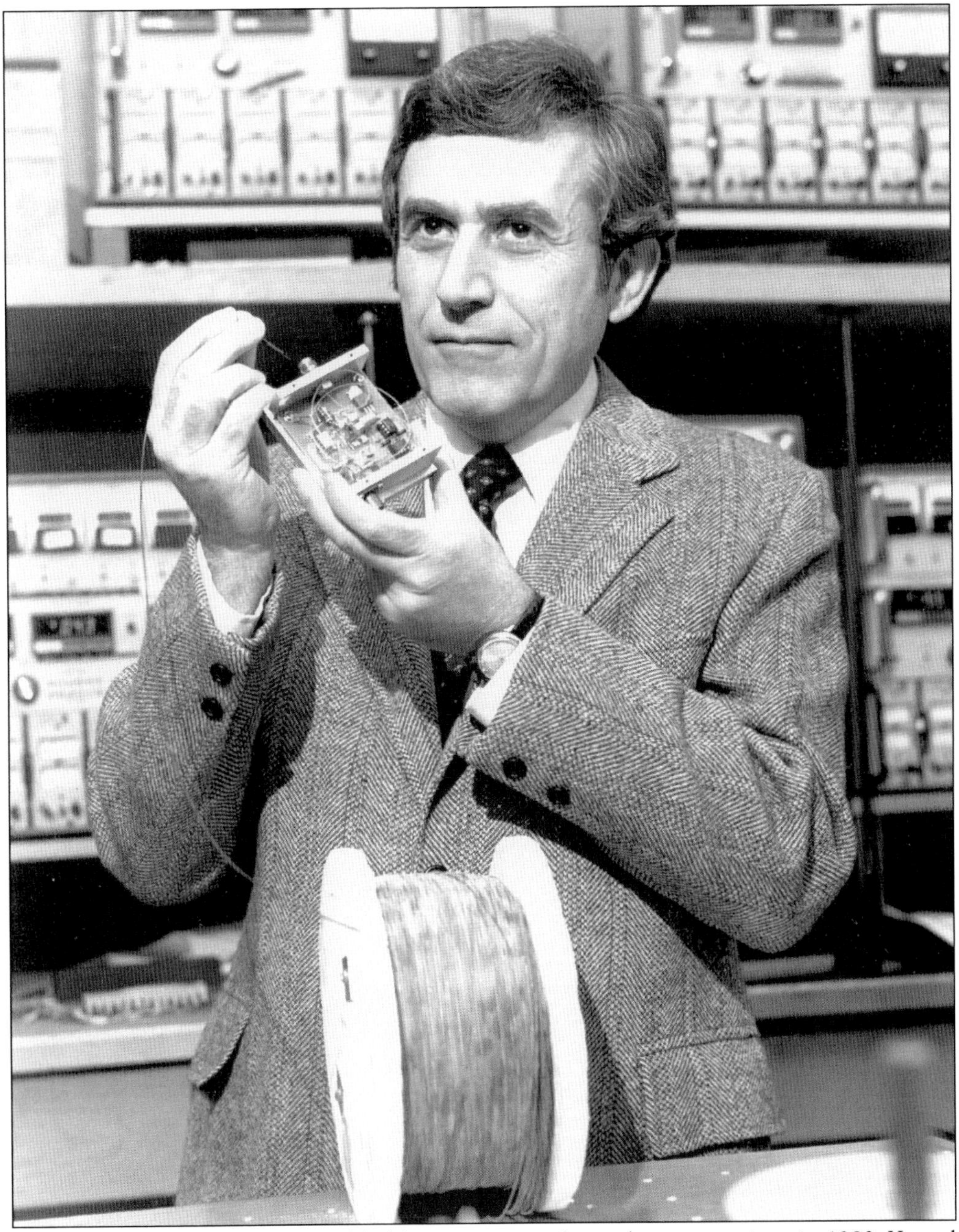

Staff vice president Henry Kressel holds up a semiconductor laser transmitter in 1980. Kressel spent his technical career reducing the size and operating temperature and increasing the durability of microscopic semiconductor lasers, efforts essential to their practical use in fiberoptic networks. A partner and a senior managing director at Warburg Pincus, Kressel became chairman of the board of directors of Sarnoff Corporation in July 2001.

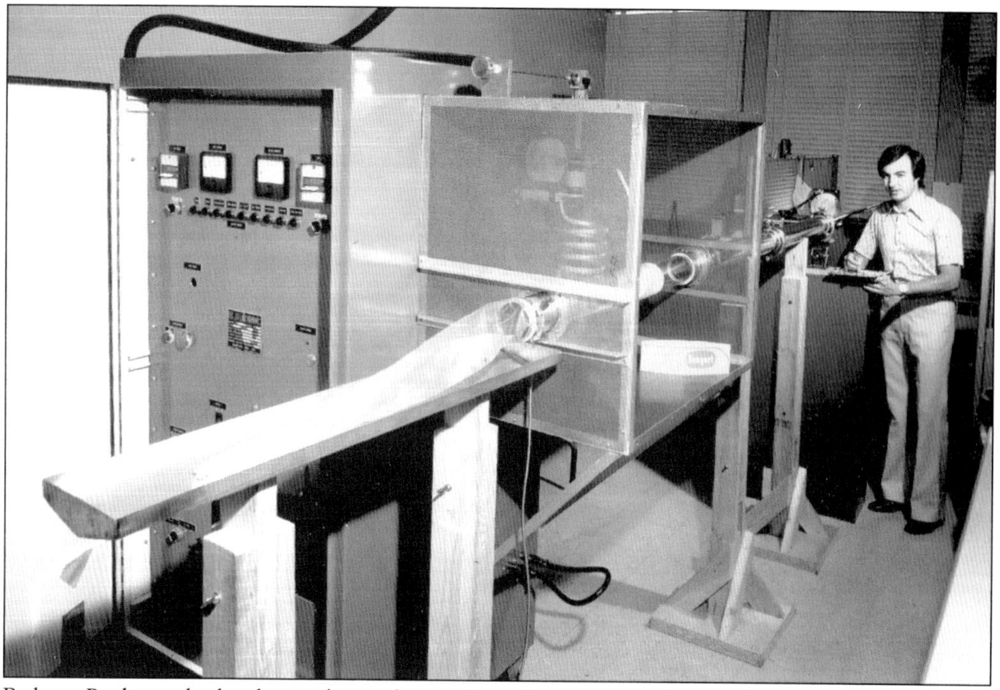

Robert Paglione looks down the turkey roll cooker that he designed using radio-frequency heating and delivered to RCA's subsidiary, Banquet Foods, in early 1980.

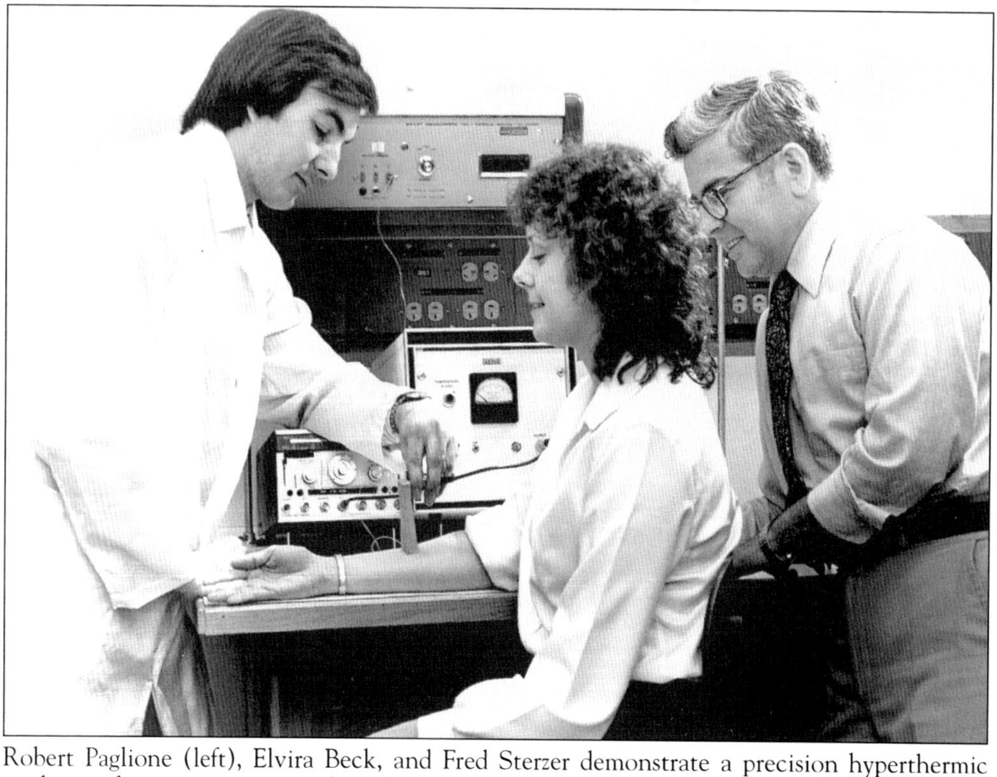

Robert Paglione (left), Elvira Beck, and Fred Sterzer demonstrate a precision hyperthermic applicator for cancer treatment based on microwaves in 1979.

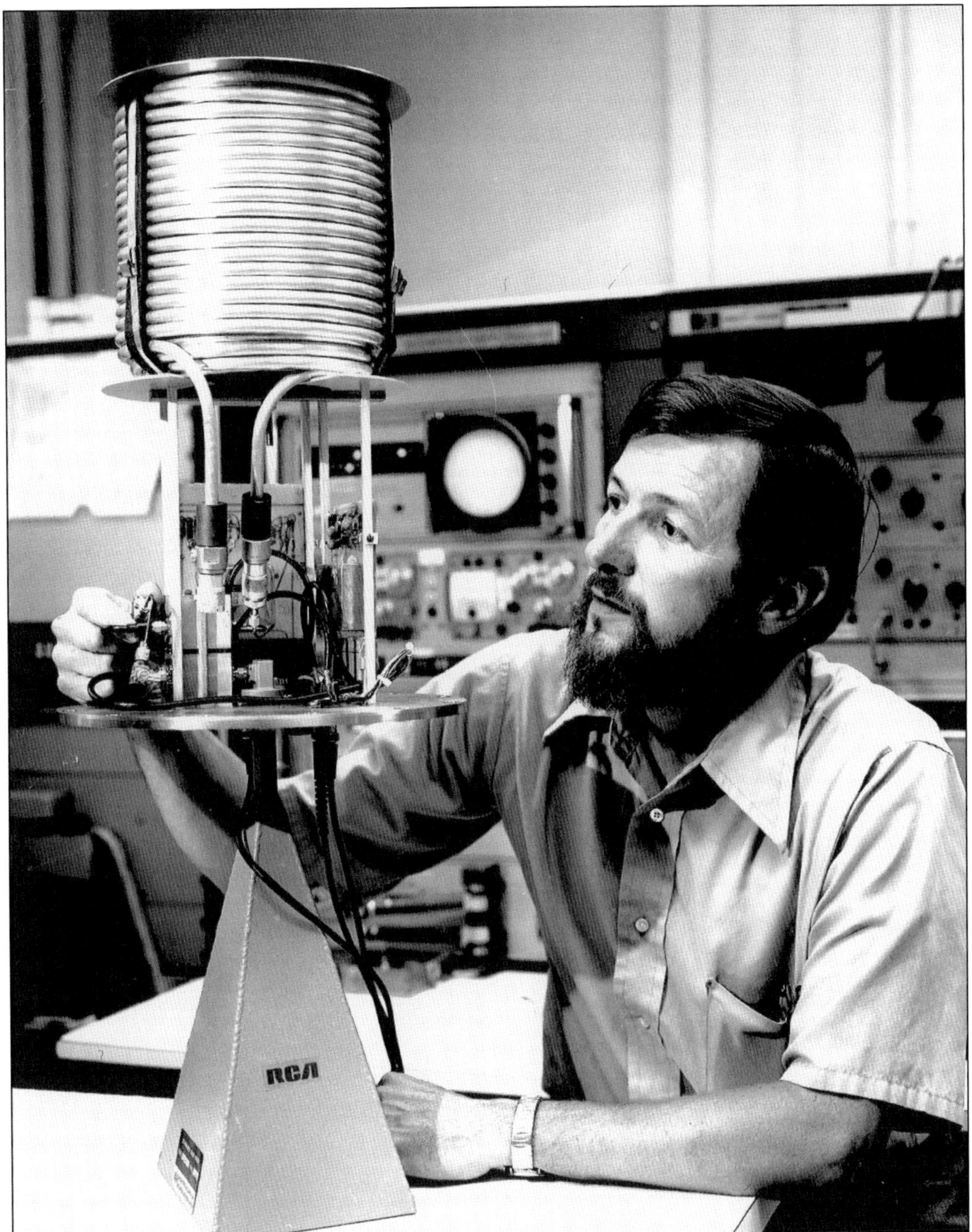

Henry Johnson of the Microwave Technology Center tweaks the microwave radar he designed for Bethlehem Steel and other companies' blast furnaces in 1980.

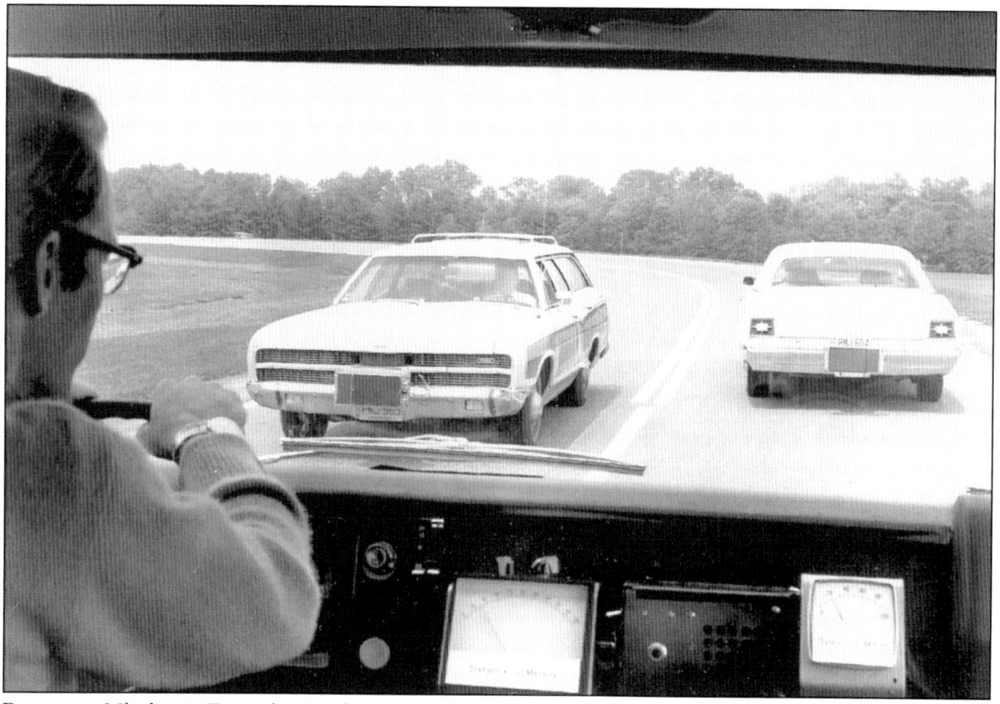

Reviving Vladimir Zworykin's ideas for using electronics to increase highway safety in 1972, the center's researchers developed a radar system using integrated circuits to sound warnings or apply the brakes when cars become too close to one another.

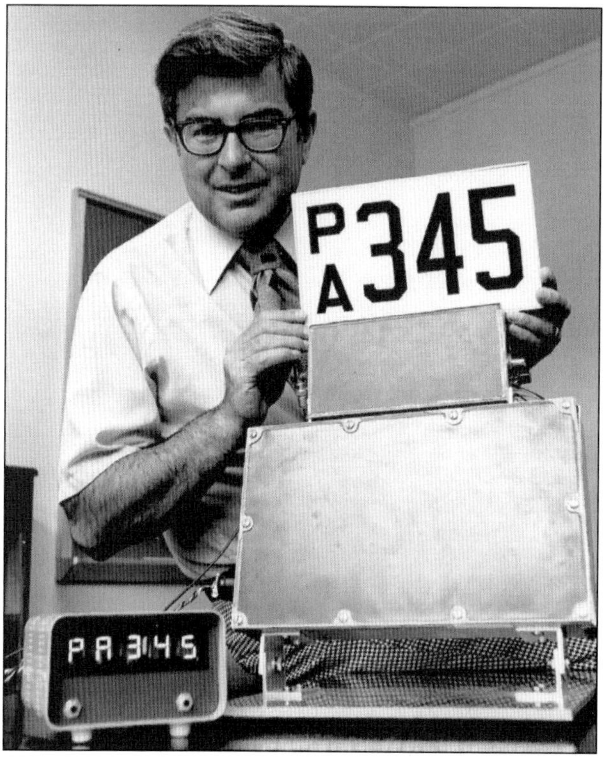

Fred Sterzer, director of the Microwave Applied Research Laboratory, shows off in July 1974 an electronic license plate containing microwave communications circuitry to improve traffic safety, reduce auto theft, and automate tolls and other services.

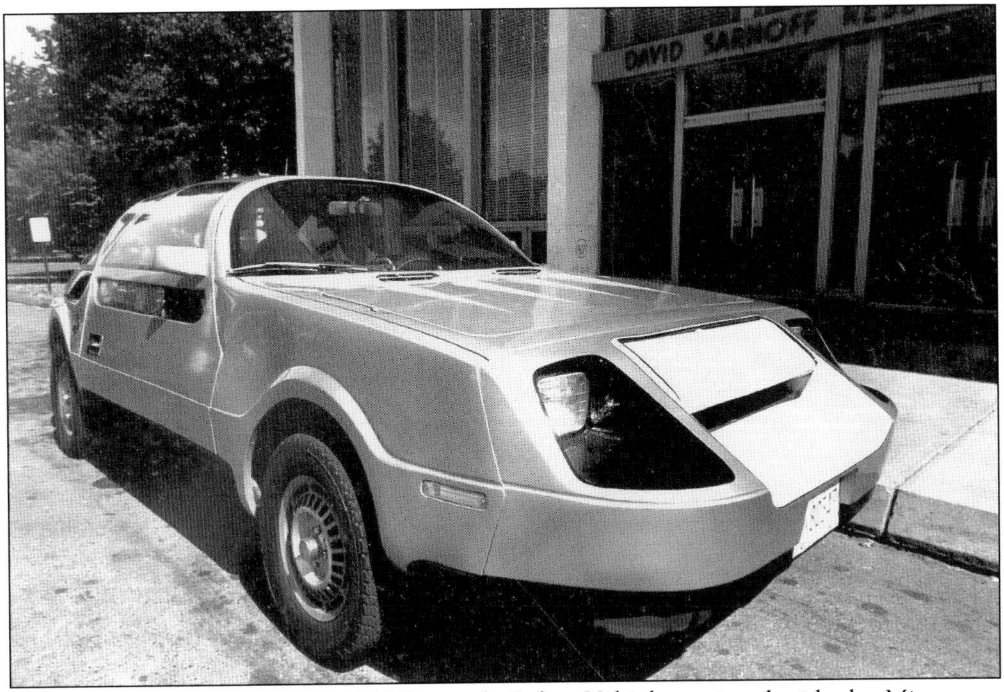

Pictured here is the Minicars 1976 Research Safety Vehicle equipped with the Microwave Applied Research Laboratory's radome in the front.

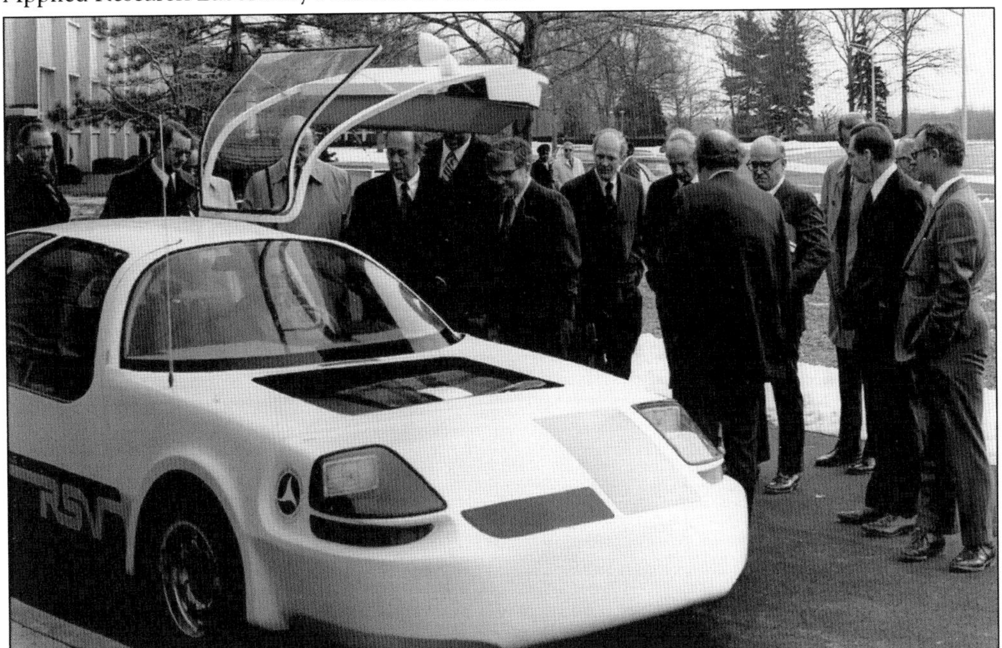

RCA's board member Cecile Selby sits in the driver's seat of the Research Safety Vehicle in 1976 while Fred Sterzer looks in. At the far left is director and future RCA chairman Edgar Griffiths. On Sterzer's left is director and future chairman Thornton Bradshaw; to his right are director William Hittinger, RCA Laboratories vice president William Webster, RCA president Anthony Conrad, Gerald Herzog, and Thomas Stanley, both of RCA Labs.

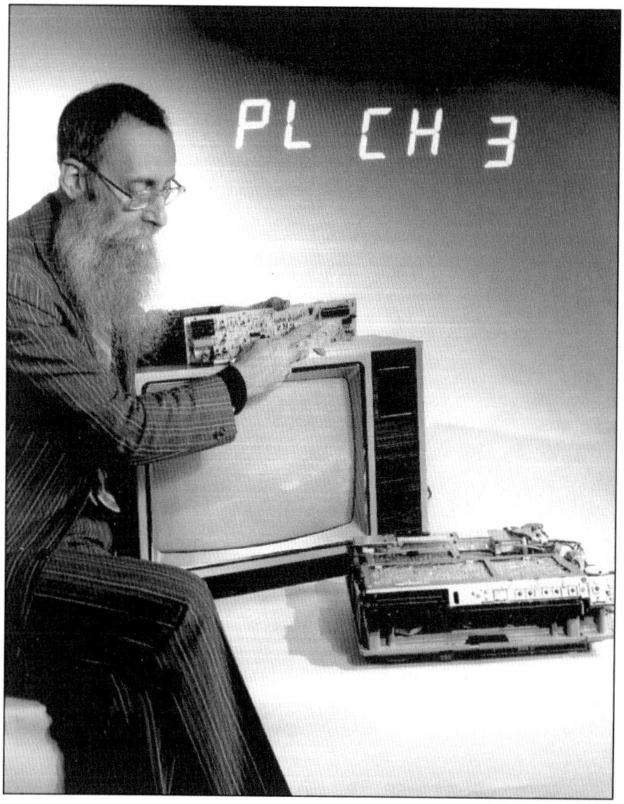

Charlie Wine demonstrates the beginning of on-screen instructions for RCA televisions with a channel identifier for components, in this case for the VideoDisc system, in 1982.

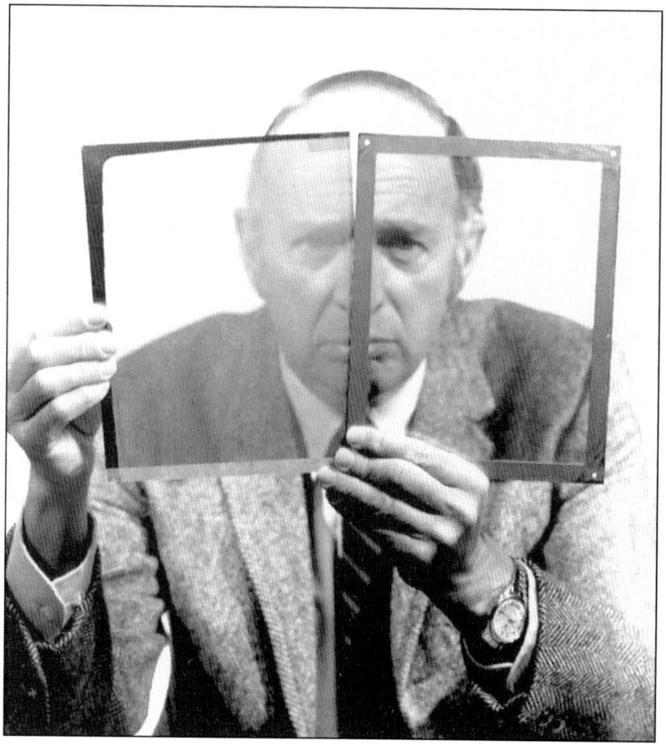

One of the problems with the shadow mask for color televisions is its limitation on the passage of electrons that can hit the screen's color phosphors, reducing the brightness of the picture. In 1982, Stanley Bloom compares a traditional shadow mask (left) to a focus mask that he invented with Carmen Catanese.

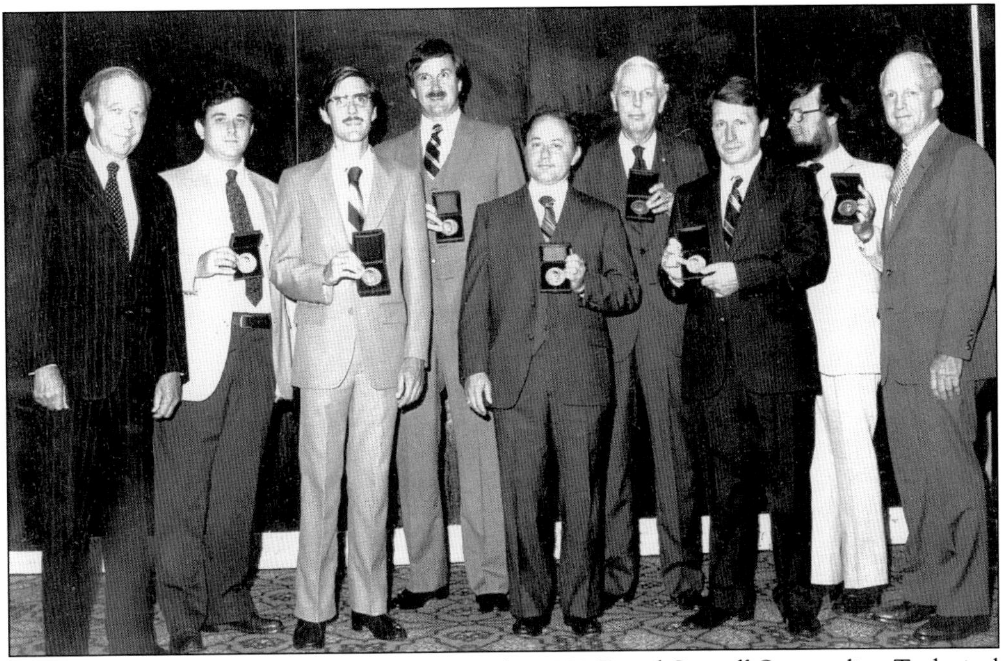

RCA chairman Thornton Bradshaw presents the 1981 David Sarnoff Outstanding Technical Achievement medal to staff members from RCA Laboratories and Consumer Electronics Division for developing the first CCD comb filter integrated circuit in color television sets. From left to right are Thornton Bradshaw, Peter Levine, Donald Sauer, James Carnes, Jack Fuhrer, Dalton Pritchard, Walter Kosonocky, William Lagoni, and William Hittinger, executive vice president for research and engineering.

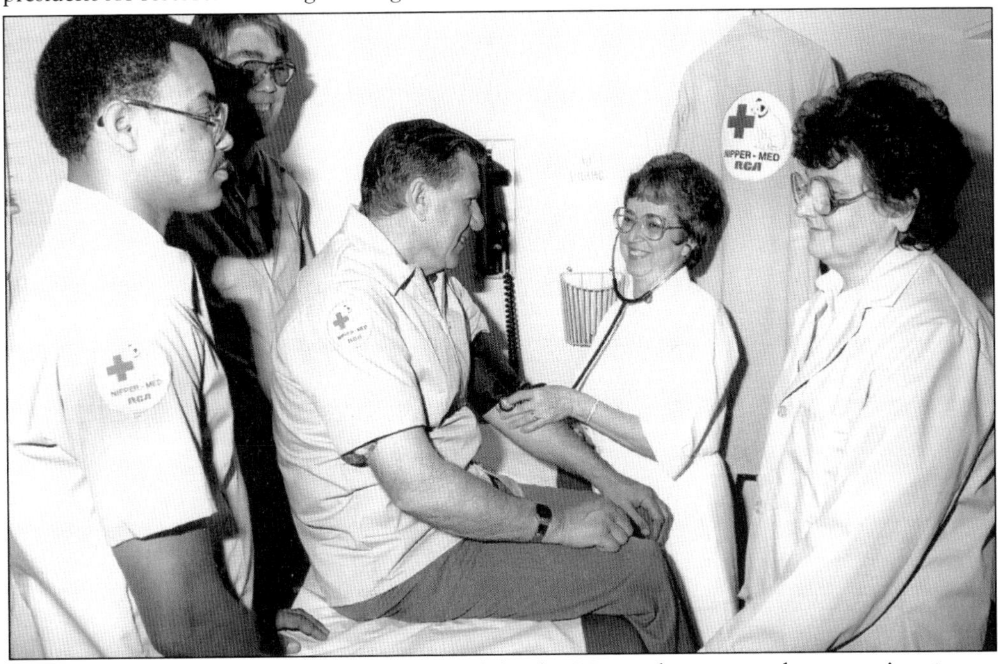

In 1978, RCA revived the 1901 trademark of the dog Nipper listening to his master's voice on a gramophone, and it was promptly adopted by a volunteer medical response team.

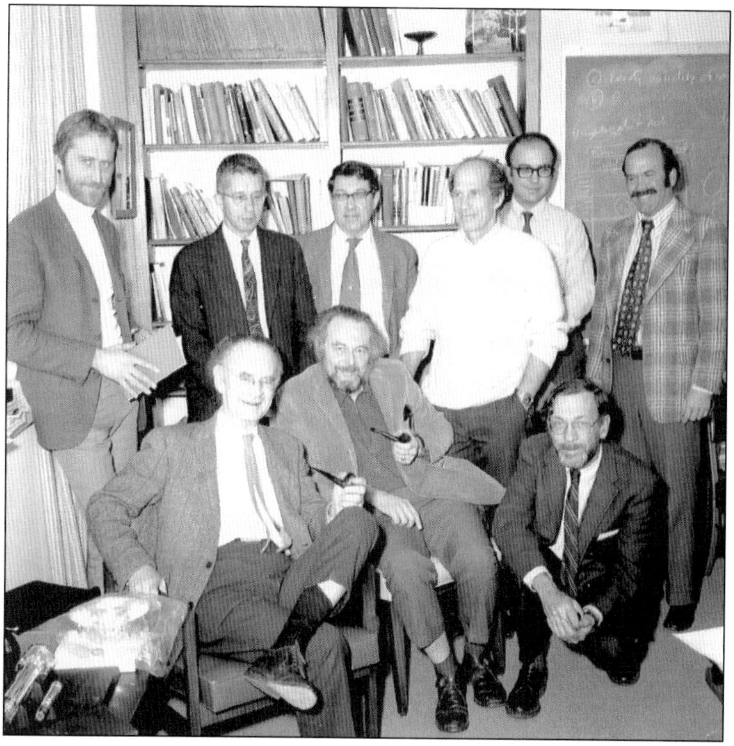

Gathered in director George Cody's office to honor Al Rose on his retirement in April 1975 are some of the group heads and fellows of the technical staff in the Physical Electronics Laboratory. From left to right are the following: (front row) Al Rose, Arnold Moore, and George Cody; (back row) Istvan Gorog, Dick Williams, Peter Wojtowicz, Ben Abeles, Roger Cohen, and Brown Williams.

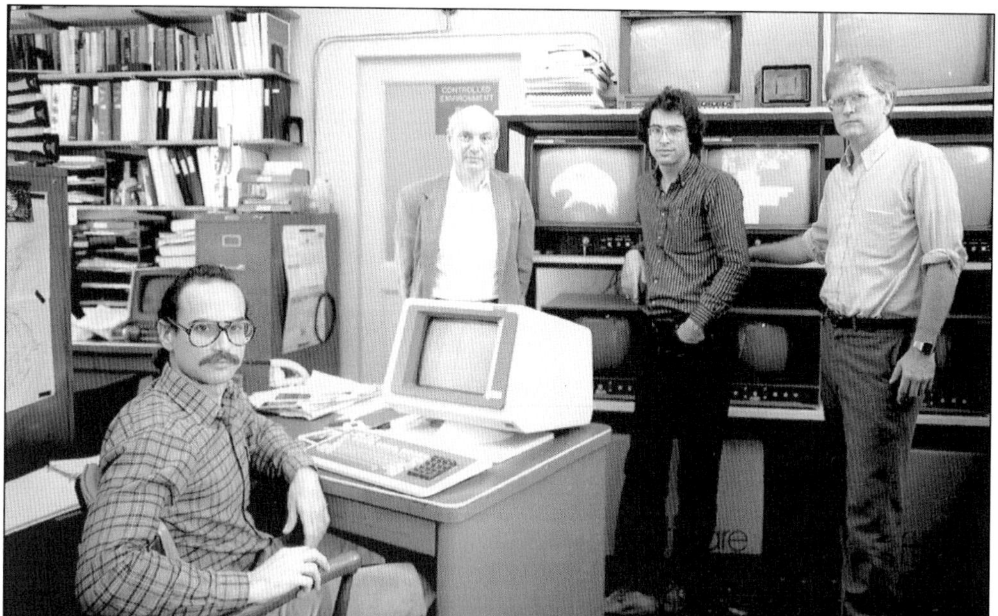

Combining the center's traditional strengths in information capture and information processing under Information Systems Research director Curtis Carlson are, from left to right, Jim Bergen, Charles Anderson, Edward Adelson, and group head Peter Burt for Advanced Image Processing. With Joan Ogden, they invented the image pyramid approach to computer vision in 1984.

Five

To a Renewed World 1987–2002

After a year of uncertainty, the fate of the David Sarnoff Research Center was resolved. On April 1, 1987, under the initiative of center president James Tietjen, General Electric donated it to SRI International, a nonprofit research organization, as a for-profit subsidiary. The center switched from servicing one company's interests to catering to the needs of corporate clients around the world, while expanding its contract work for the government.

Over the next 15 years, the center rode the explosions in digital and biomedical technologies and the roller-coaster economy. The digitization and networking of information changed the nature of much innovation from hardware to software in form, from video research to communications to genomic processing. As the fusion of electronics and biology finally took place, the center (renamed Sarnoff Corporation in 1997) diversified appropriately. The surge of venture capital during the 1990s fueled the spin-off of 21 companies formed to commercialize Sarnoff's inventions.

Sarnoff's staff's skills, along with the substantial portfolio of patents and collection of specialized products, have turned the company from a research-and-development organization to an innovation services provider, responding to the needs of varied customers and markets along the S-curve that takes an idea to a useful and popular reality.

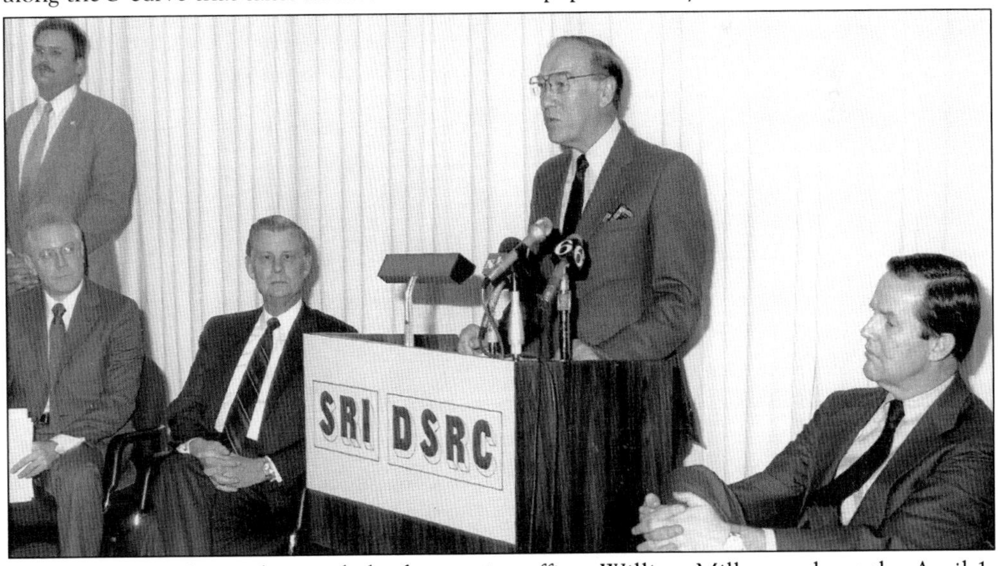

SRI International president and chief executive officer William Miller speaks at the April 1, 1987, press conference announcing the center's donation by General Electric to SRI. Seated in the center's lounge are, from left to right, James J. Tietjen (president and chief executive officer of the center), Edward Hood (General Electric vice chairman), and Thomas Kean (governor of New Jersey).

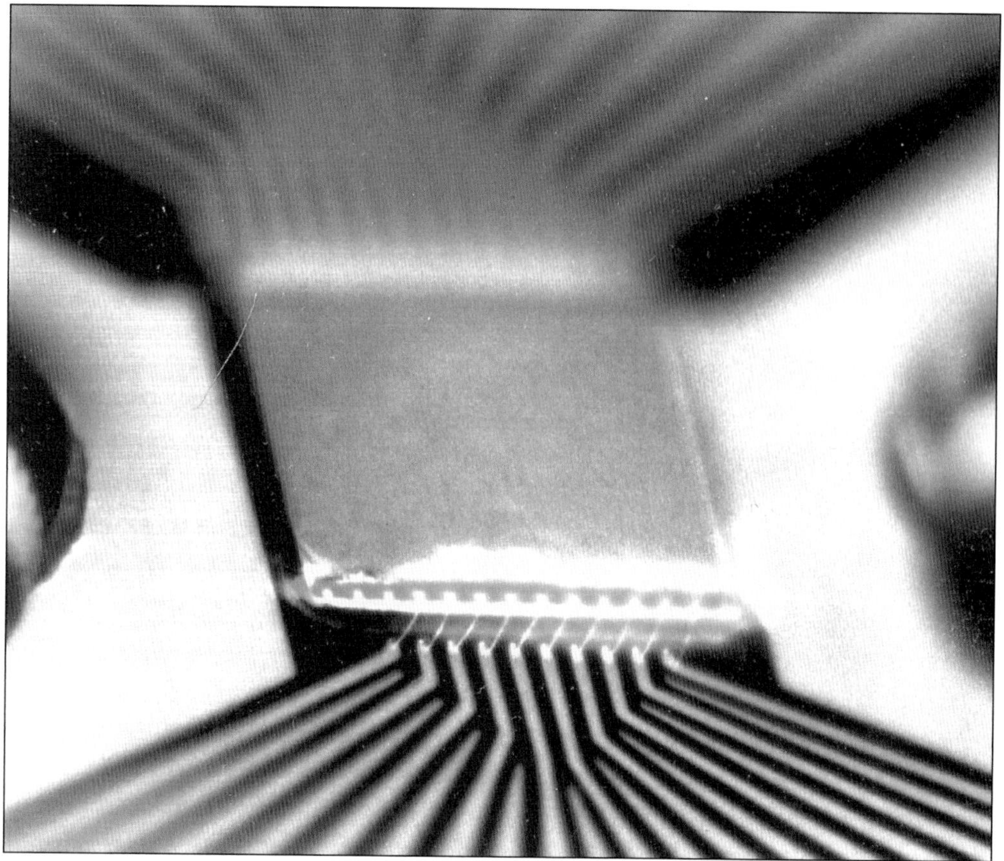

One of the center's first innovations was this Grating Surface Emitting Array laser diode chip in 1988. Optoelectronics director Michael Ettenberg led the team that developed the GSE Array to enhance fiberoptic communications through its simple and rugged fabrication, unfaceted and highly coherent beams, and placement of 100 switchable lasers on the surface of an integrated circuit wafer.

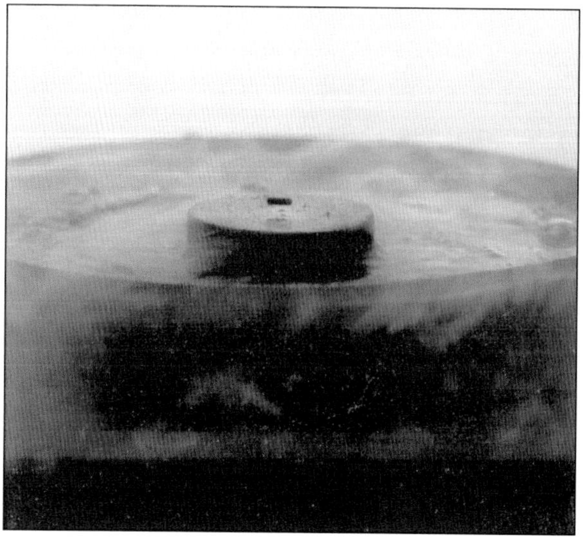

In 1988, James Matey's levitating magnet over a superconductor cooled by liquid nitrogen continued innovation at the center on new materials and processes for superconductors that operate closer to room temperature.

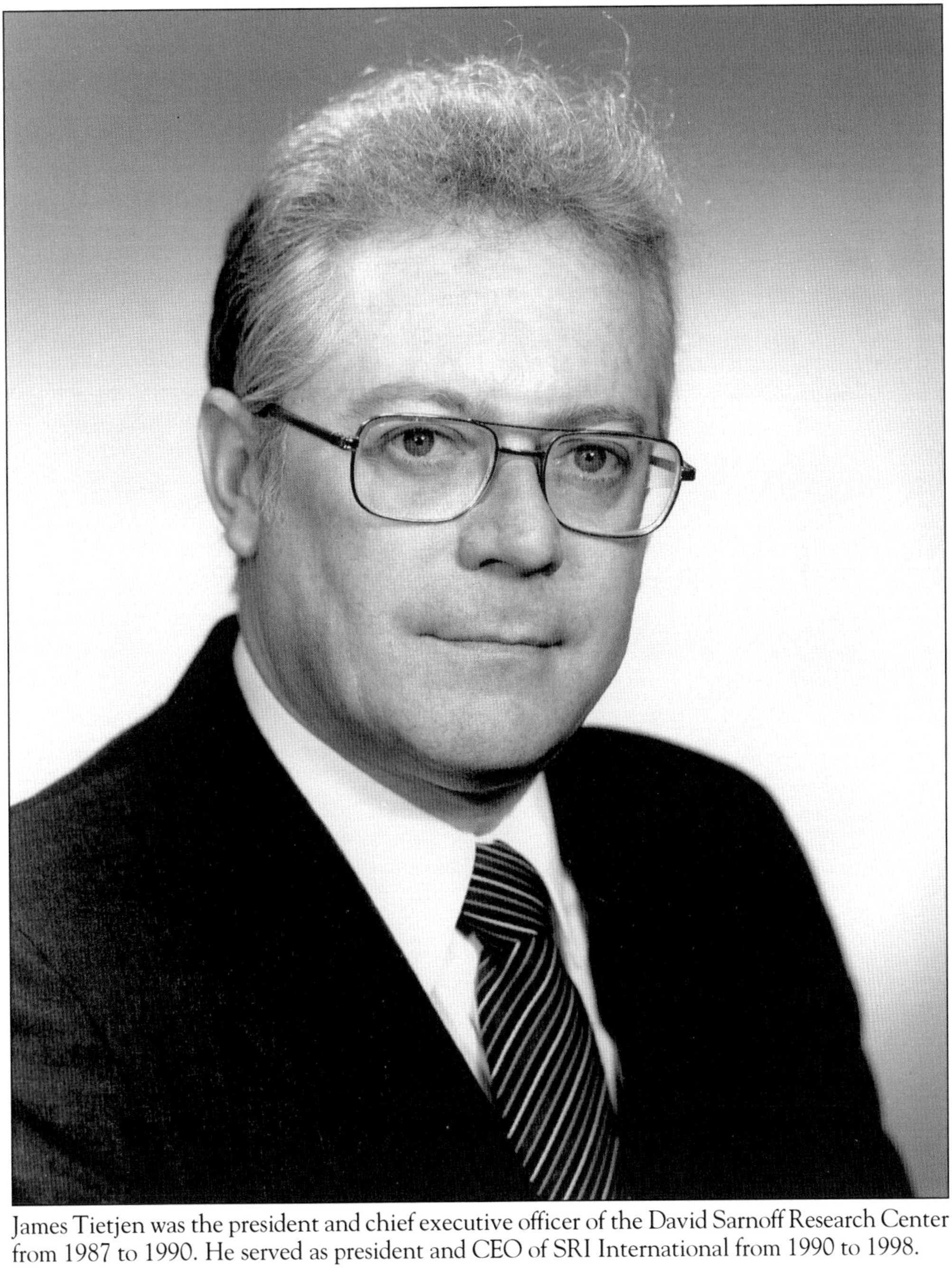

James Tietjen was the president and chief executive officer of the David Sarnoff Research Center from 1987 to 1990. He served as president and CEO of SRI International from 1990 to 1998.

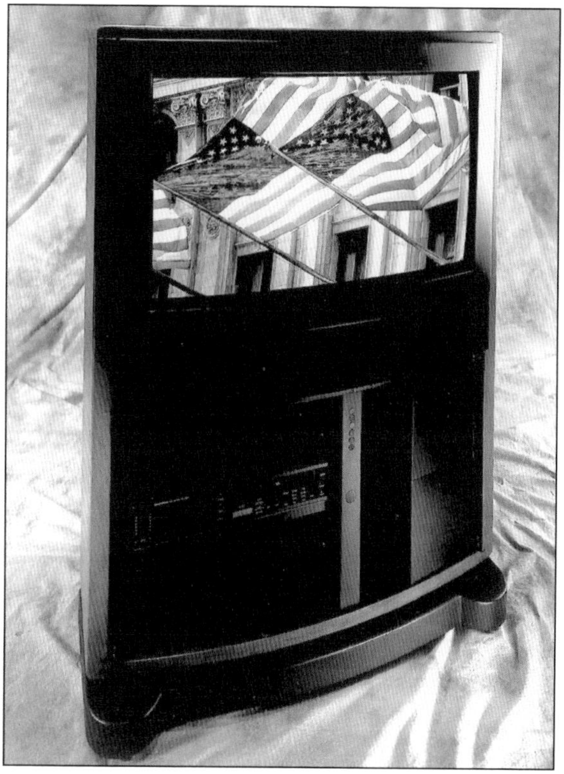

Under Curtis Carlson's leadership, the center was a member of the Grand Alliance that developed the American high-definition television standard, established in 1996 after nine years of technical innovations, political debate, and business discussions. Sarnoff staff contributed to the compression format fitting an HDTV signal into the same channel bandwidth used by the analog standard, built the analog-to-digital encoder, and provided the systems integration and evaluation, from which 9 of 10 recommendations were adopted by the Federal Communications Commission.

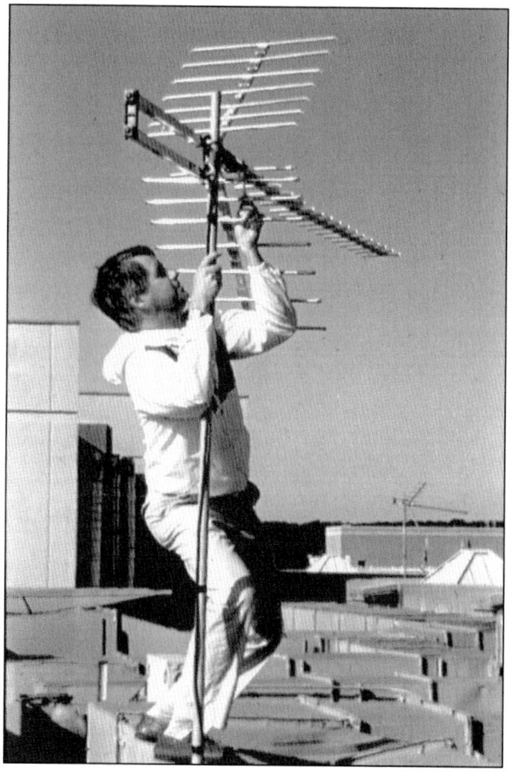

Manpower complemented brainpower in the development of HDTV. Center engineer Frank Lang shinnies up an HDTV antenna in Washington, D.C., to adjust it during tests.

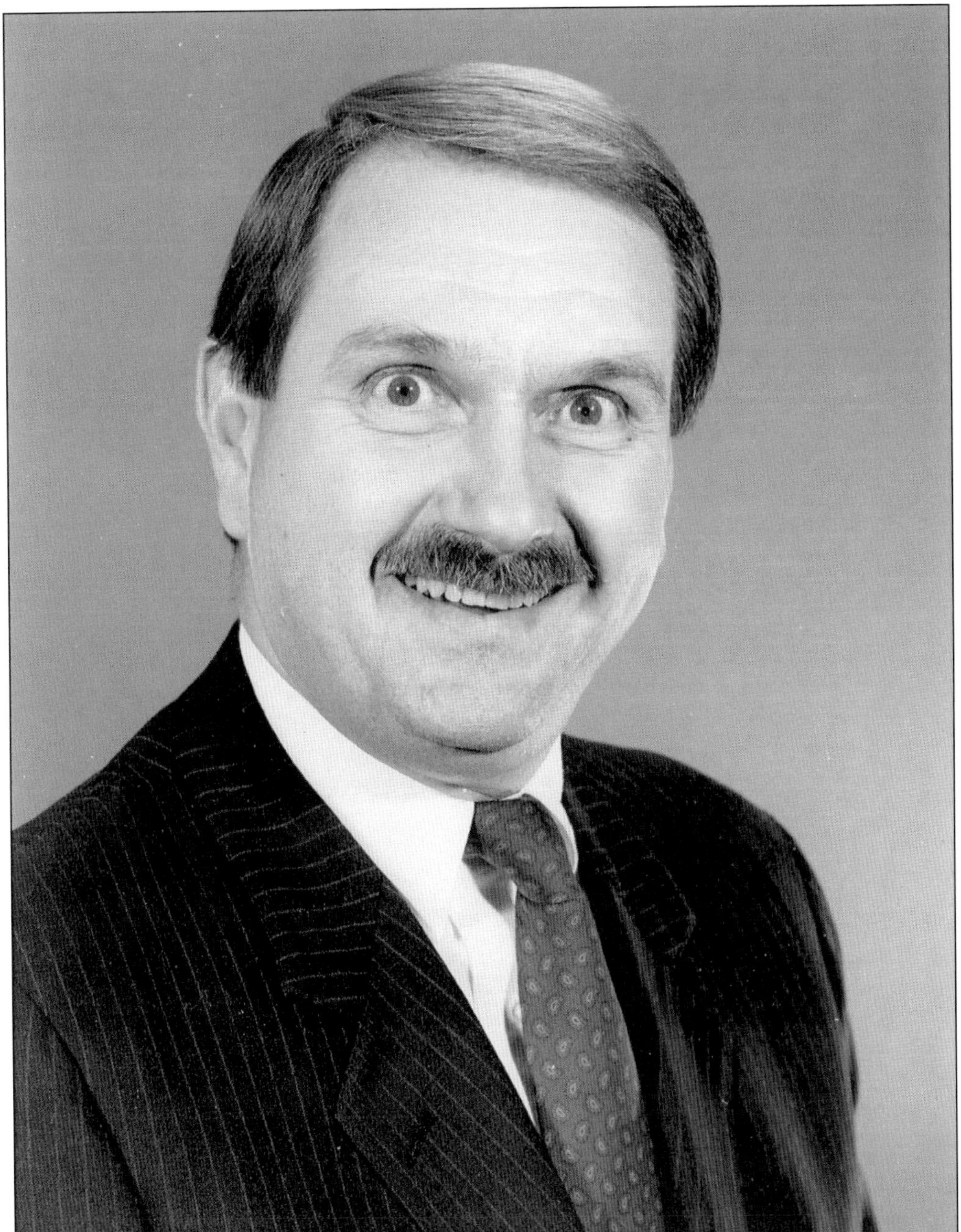

James Carnes served as president and chief executive officer of the David Sarnoff Research Center (1990–1997) and of Sarnoff Corporation (1997–2002).

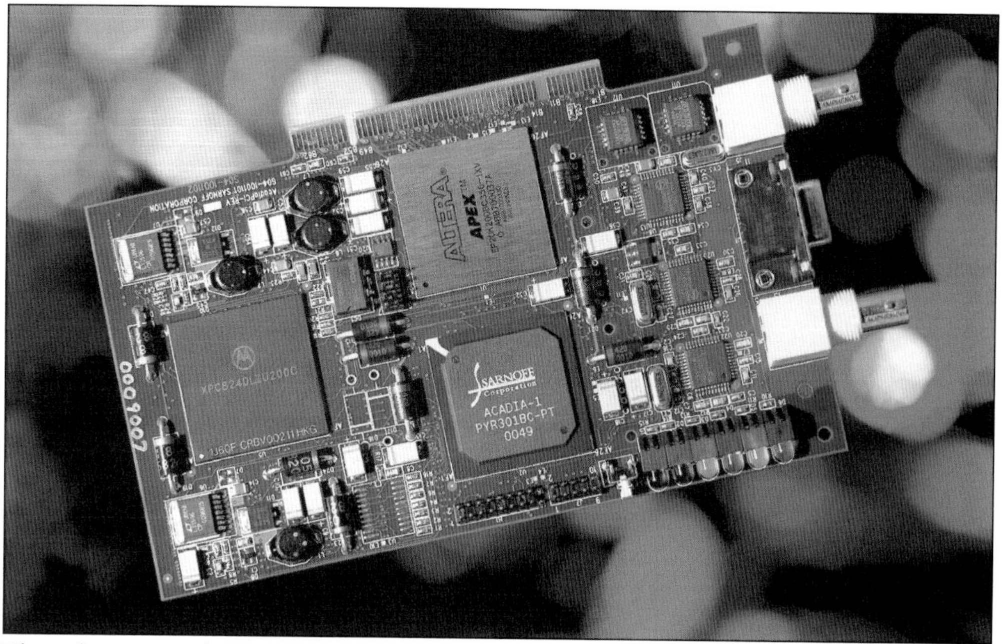

The culmination of over 25 years of work in pyramid vision processing at Sarnoff, the Acadia I vision processing chip, introduced in 2000, offers real-time video processing for consumer products and for desktop and laptop computers.

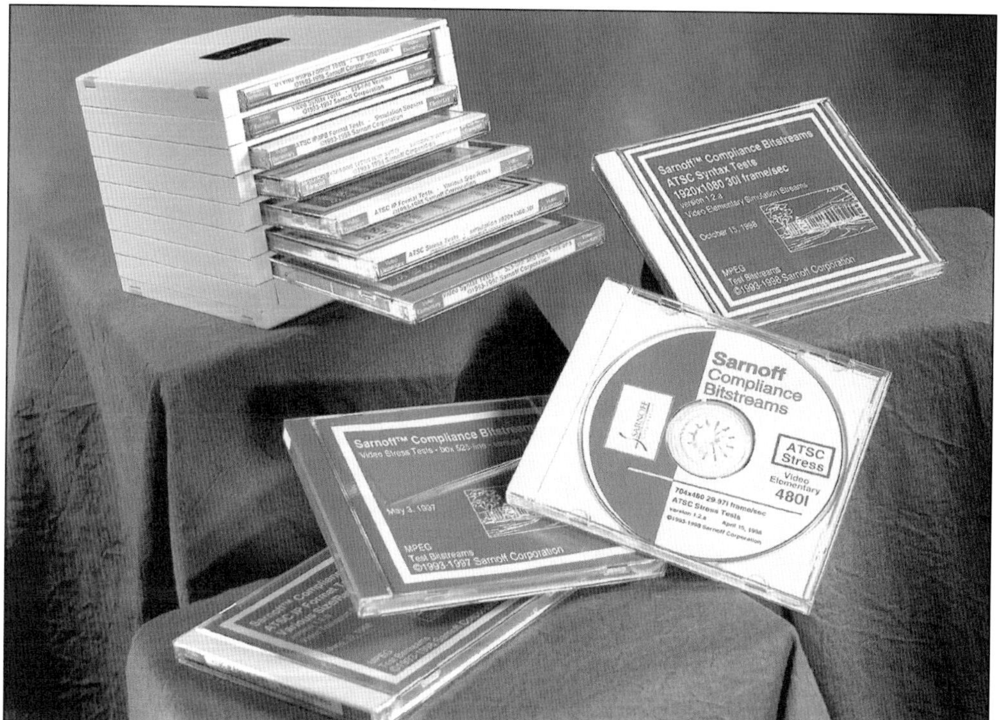

Sarnoff's Emmy-winning Compliance Bitstreams software provides a new set of tools for evaluating the MPEG-based video decoders used by broadcasters as they make the transition to digital television.

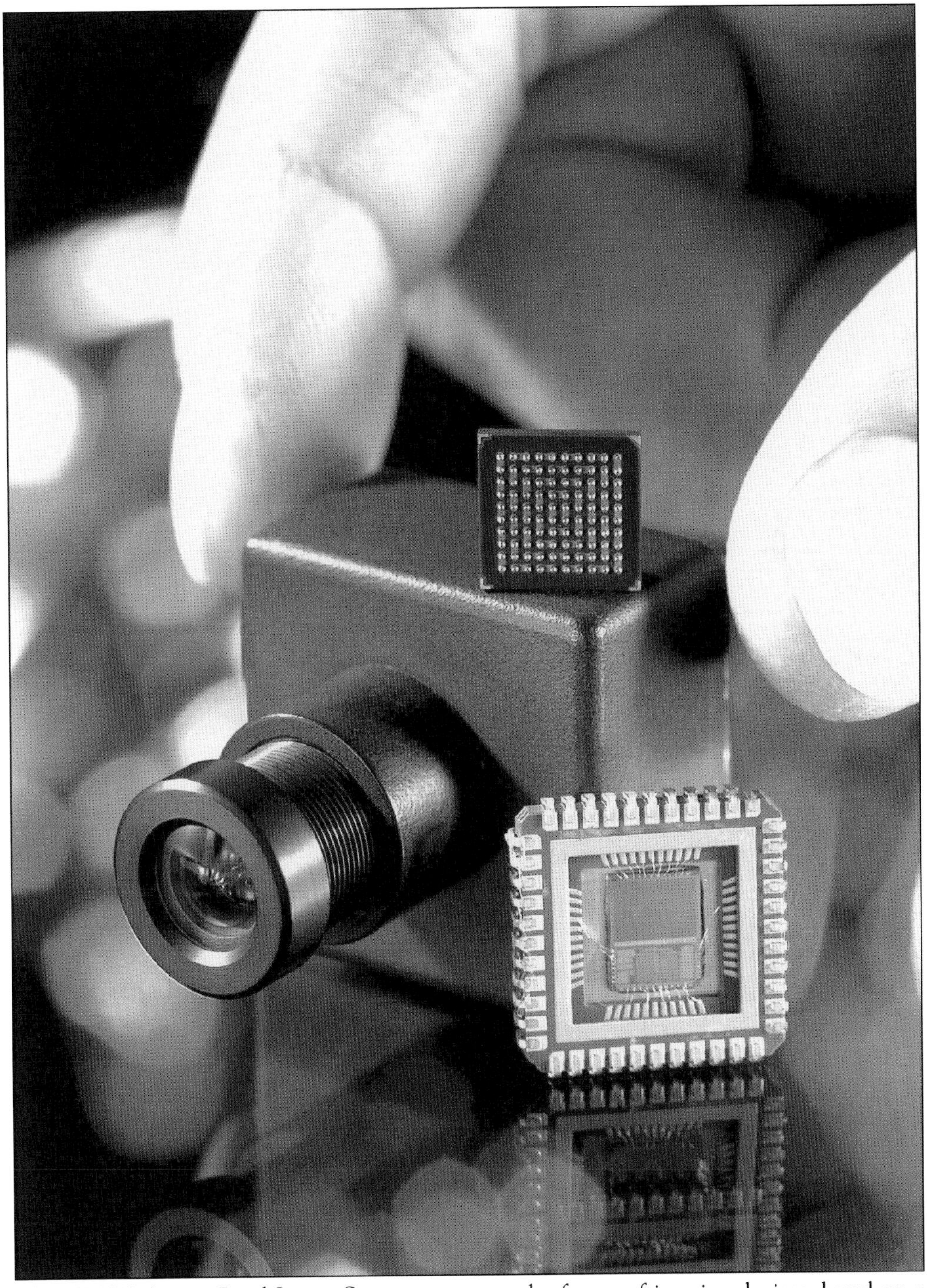

This CMOS Active Pixel Image Sensor represents the future of imaging devices, based on a transistor technology innovated at the David Sarnoff Research Center in the 1960s. Sarnoff's latest camera provides a signal-to-noise ratio of 110 decibels, compared to 50–60 decibels for most CCD cameras, equivalent to that in photographic film.

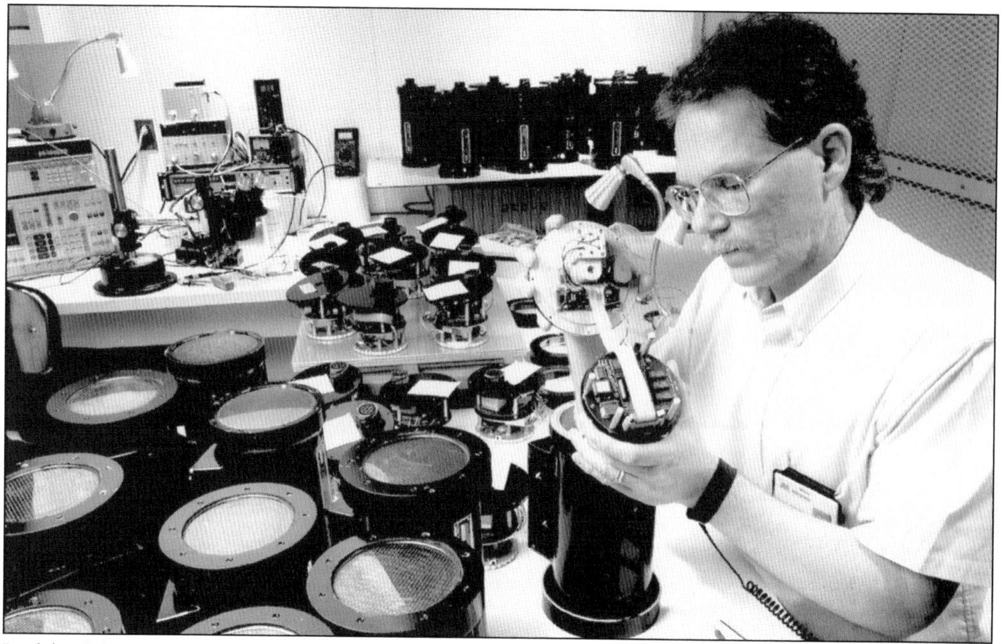

Led by Microwave Systems director Ray Camisa, Sarnoff has built and sold over 1,500 True Ground Speed Sensors, which solve the problem of wheel slippage on railroads. These sensors use microprocessors and microwave radar to compare ground speed with wheel speed and allow the locomotive's motor-control system to adjust accordingly.

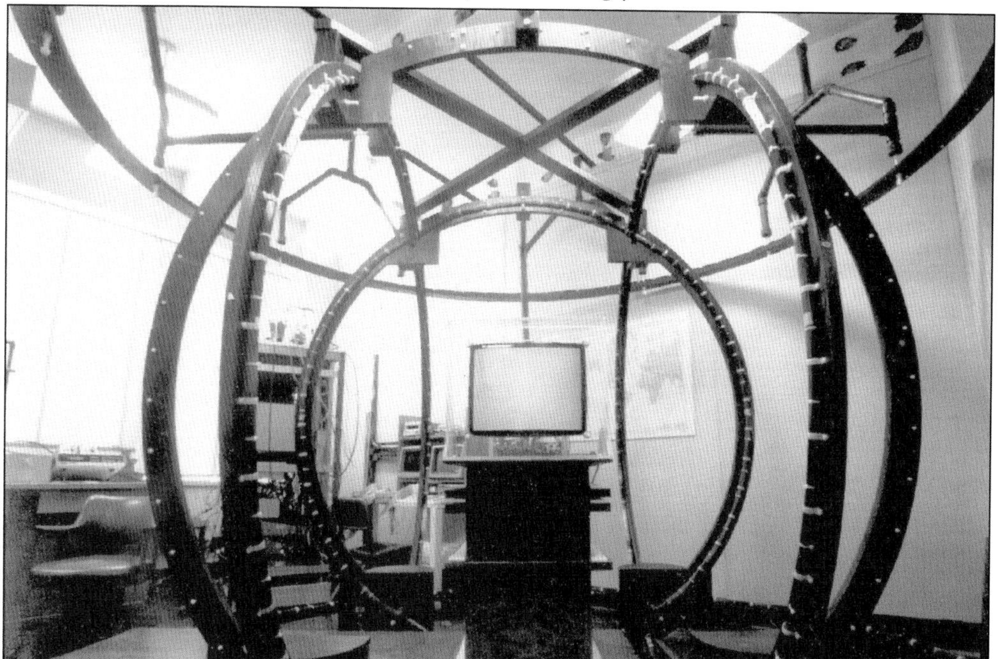

During the 1990s, part of Harry Olson's free-field acoustics room was turned into an electromagnetic test bed. These Helmholtz coils of wood and copper wire, named after the 19th-century German physicist, provide a constant magnetic field that can be adjusted to simulate that of the earth anywhere on the planet and test its effect on television sets.

After over 40 years of experience in integrated circuitry innovation, Sarnoff's staff continues to offer a wide array of cutting-edge skills, services, and products in the design, development, and fabrication of integrated circuits and systems on a chip.

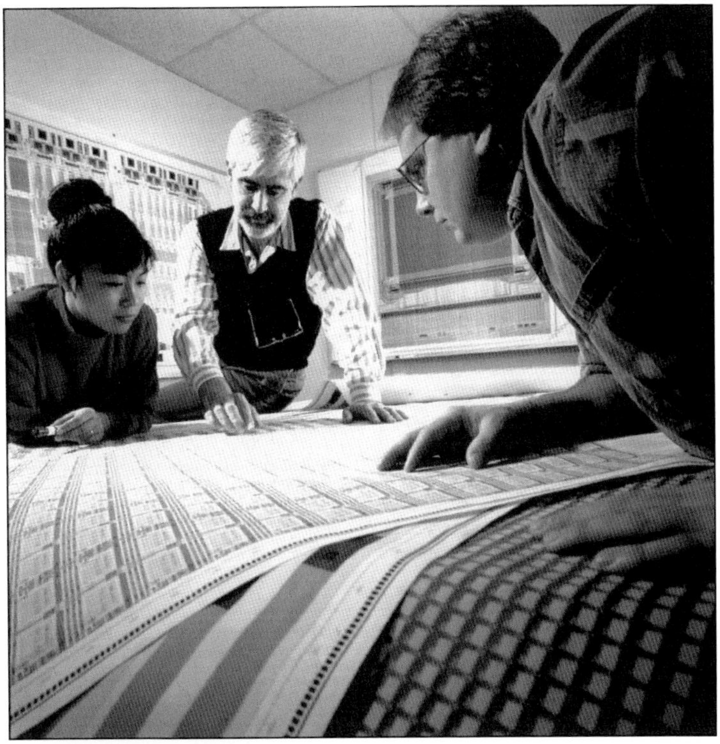

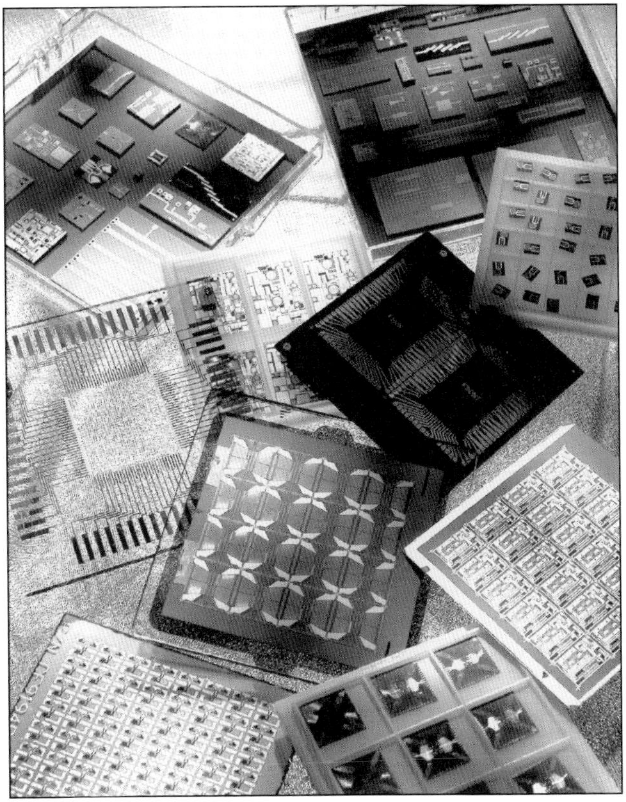

The organization that invented thin-film transistors continues to innovate the technology and its fabrication, offering a wide variety of designs and services.

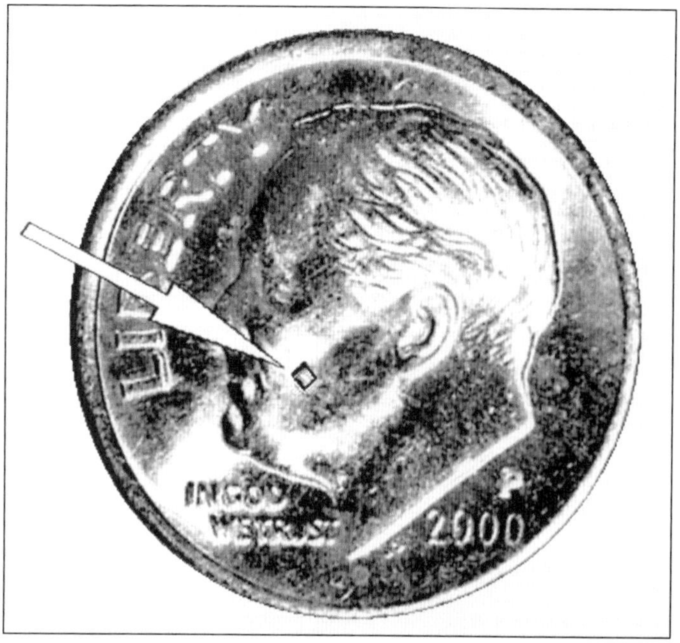

Pictured here is one of Sarnoff's radio frequency identification tags, which can lower the cost of product and process tracking, access control, and financial transactions.

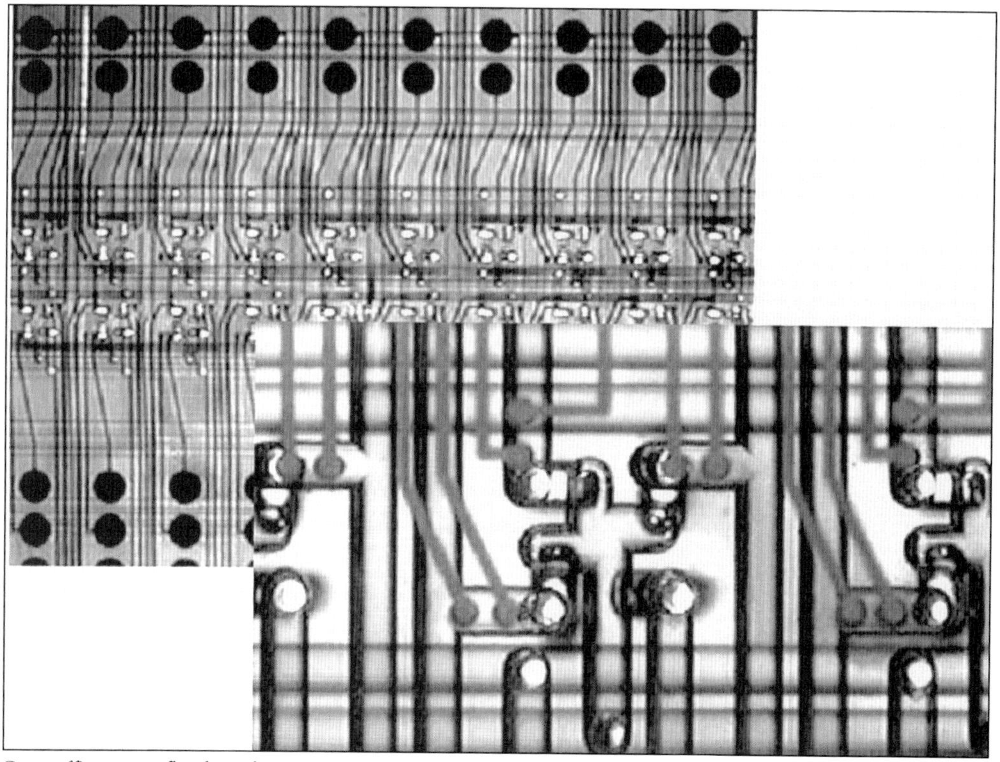

Sarnoff's microfluidic chip, introduced in 1995, provided the first integrated microfluidic detection system for biochemical assays, or "lab on a chip." Based on pioneering patents in the electrohydrodynamic pumping of fluids by Sterling McBride, Satyam Cherukuri, and Pamela York, the microfluidic chip led to the organization of Orchid BioSciences, the first Sarnoff technology venture to go public, in May 2000.

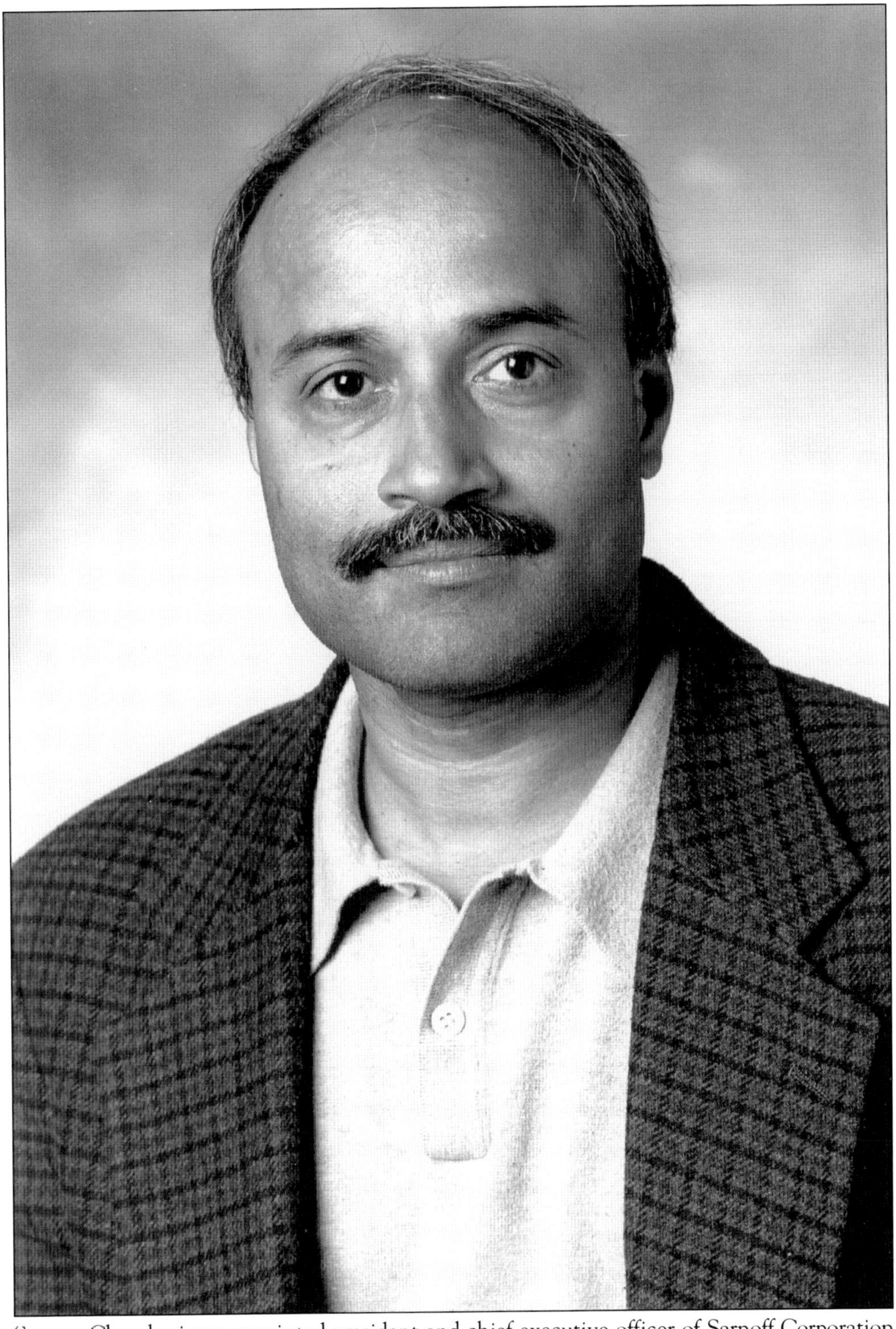

Satyam Cherukuri was appointed president and chief executive officer of Sarnoff Corporation in May 2002.

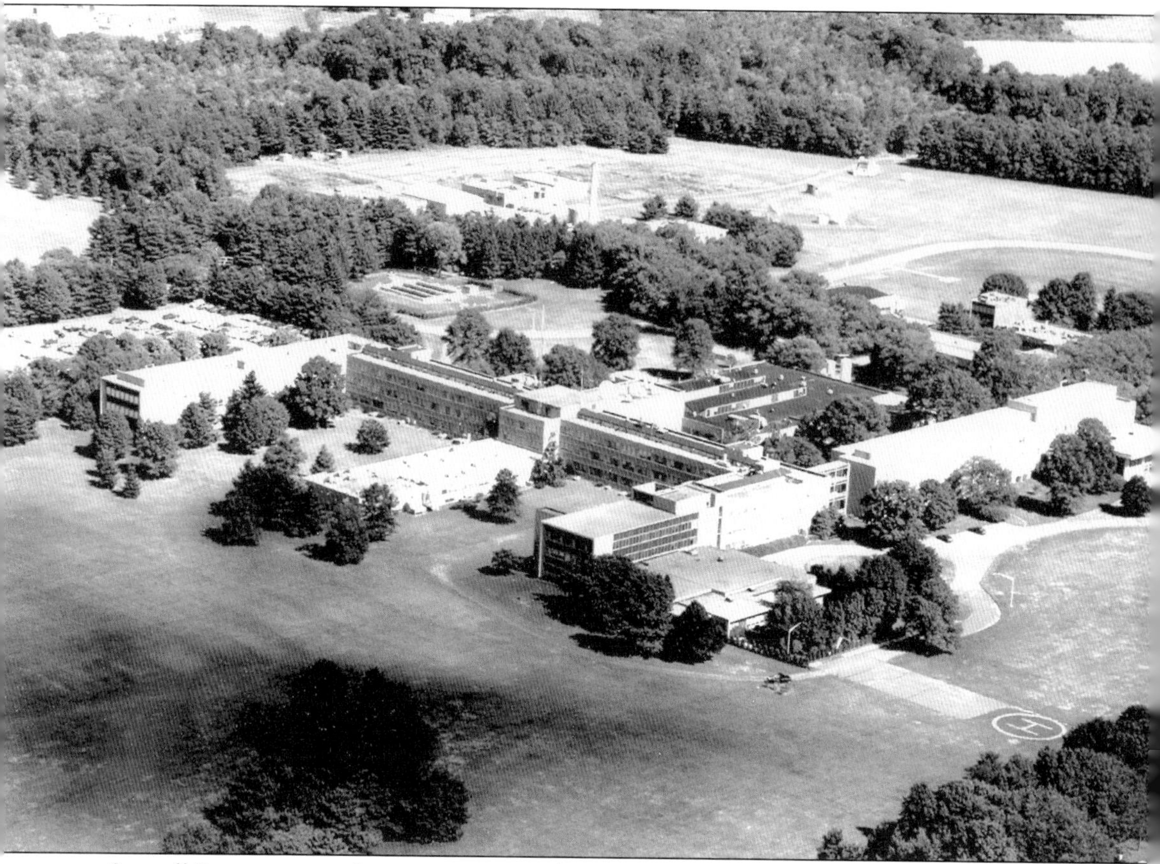
Sarnoff Corporation is shown as it appeared in 2002.